Math for Writers:

Tell a Better Story, Get Published, Make More Money

By Laura Laing

$\lim\limits_{x \to \infty}$

Limitless Press

First Printing, 2014

$$\lim_{x \to \infty}$$

Limitless Press Co.
5700 Ridgedale Road
Baltimore, MD 21209
www.mathforgrownups.com
writers@mathforgrownups.com

Quantity sales. Special discounts are available on quantity purchases by corporations, associations, and others. For details, contact the publisher at the address above.

Printed in the United States of America

ISBN 978-0-9914465-0-6

This publication is designed to provide accurate and authoritative information with regard to the subject matter covered. It is sold with the understanding that the publisher is not engaged in rendering legal, accounting, or other professional advice. If legal advice or other expert assistance is required, the services of a competent professional person should be sought.
—From a Declaration of Principles jointly adopted by a Committee of the American Bar Association and a Committee of Publishers and Associations

Many of the designations used by manufacturers and sellers to distinguish their product are claimed as trademarks. Where those designations appear in this book and Laura Laing was are of a trademark claim, the designations have been printed with initial capital letters.

Praise for Laura Laing and *Math for Writers*

"This is the age of big data. For journalists, big data means big stories. Laura Laing takes the mystery out of math that writers need to understand big data. I highly recommend her work—it just adds up."

—Michelle V. Rafter, "WordCount: Freelancing in the Digital Age"

"I used to tell people that as a writer, I was good with words, not numbers. Thanks to Laura Laing, I'm good at both, now."

—Sandra Beckwith, *Get Your Book in the News: How to Write a Press Release That Announces Your Book*

"Math? Let's just say it's not my strong suit. Laura Laing's book is invaluable, explaining how to do the math, and how it can make you a better writer and a more successful freelancer."

—Kelly James-Enger, *Six-Figure Freelancing: The Writer's Guide to Making More Money, Second Edition*

"Thanks to Laura Laing, math is no longer like the scary monster under the bed. Now we writers can easily deal with math in our business life without truckloads of antacid."

—Michele Wojciechowski, *Next Time I Move, They'll Carry Me Out in a Box*

Table of Contents

Acknowledgments

THE BEST BUSINESS ADVICE I ever got was to surround myself with smart people who are where I want to be. That's exactly what I've done throughout my freelance writing career. As a result, I've had the distinct pleasure of getting to know some of the top writers, editors, and publicity experts in the country. Each has played a very important role in my work, helping me get where I am today.

To honor these fantastic folks, I've included them as "characters" in this book. You'll see variations of their names in the math examples. I thought I'd introduce them here.

Linda Formichelli was the first real, live freelance writer I ever met. We connected on a long-since-gone online forum in 2001. She proofed my first query to *Parents* magazine and squealed with delight when the editor bought that idea. A year or so later, she published *The Renegade Writer* with Diana Burrell. Since then, she's been advising budding freelance writers through her website, online courses and more books, like *Write Your Way Out of the Rat Race… and Step Into a Career You Love.*

Jennifer Lawler is the reason I've written this book. A while back, I took her book proposal writing course, developing an idea for a math book for parents. It never sold. But Jennifer didn't forget. As an editor at Adams Media, she recommended me, when the editing team was looking for a writer for *Math for Grownups*, my first book. I've been writing about math since then. Jennifer served as the development editor for this book. Check out her own titles, including *Dojo Wisdom for Writers*.

Michele "Wojo" Wojciechowski is funny. She's also one of the only freelance writers I know from my area—and one of the only writers I have lunch with. Each year, Wojo and I take a bus ride to New York City to attend the American Society of Journalists and Authors (ASJA) conference. Her first book, *Next Time I Move, They'll Carry Me Out in a Box* is hilarious. So is her stand-up comedy.

One of the first freelance business books I read was by **Kelly James-Enger**, called *$ix Figure Freelancing*. I finally got to meet her at an ASJA conference three years ago, and it was like we were long-lost friends. These days, she's running her own small publishing house, just for writers and hobbyists-turned-entrepreneurs, called Improvise Press.

Michelle Rafter is a blogger extraordinaire. WordCount is a freelance writer's dream—offering up-to-the-moment advice on how to use technology and social media to its fullest. Michelle is also the founder of the WordCount Blogathon, which encourages freelance writers to blog for 30 days straight each June.

One of the first things I did after writing *Math for Grownups* was to take **Sandy Beckwith**'s Build Book Buzz Book Publicity 101 e-course. Even with a background in publicity and media, I learned so much. These days, Sandy and I co-edit ASJA's blog, The Word. She's the yin to my yang. (In other words, she does the detail work, and I scramble around at the last minute to get posts ready).

Jenny Fink is another great friend from ASJA, and we also hang out at the freelance writer's water cooler—Facebook. The mother of four rambunctious boys, she's the founder of BuildingBoys.net, which offers amazing advice for parents, teachers, and those of us who think boys are pretty cool. Jenny is a shining example of how to turn one's life into an honest and helpful brand.

Jennifer Nelson got me all riled up with her book, *Airbrushed Nation: The Lure and Loathing of Women's Magazines*. In essence, she turned her long experience writing for women's magazines into an exposé of the industry, considering how these publications dictate the definition of beauty. It was a bold and impressive step.

When I needed help figuring out my next steps last fall, I called **Emma Johnson**, a personal finance writer who is turning her freelance writing business on its head. Her blog, Wealthy Single Mommy, is an honest and revealing take on raising kids, dating and making money. I will treasure the enriching conversations we've had about making a living and writing. Oh, her couch is very comfortable and her kids are cute, too.

Kayt Sukel is someone who doesn't hold back. It's my favorite thing about her. Her book, *Dirty Minds* or *This Is Your Brain on Sex*, is an uncompromising examination of the neuroscience of sex. Her next book considers the neuroscience of risk taking. Geeking out with Kayt is one of my favorite pastimes.

Gwen Moran is another straight-shooting businesswoman and writer. With credits in some of the top business publications in the world and having ghostwritten books for names you'd recognize, she's got it going on. In fact, she's sharing her expertise with small business owners at her newest venture: Bizversity. She's also one of the most generous people I've ever met, eager to share some advice or just listen to a sad story or two.

Another funny writer, **Jen Singer**, is one of the first mom bloggers I ever met. MommaSaid is a hilarious look at parenting that took a serious turn a few years ago, when Jen was diagnosed with cancer. Undaunted, she incorporated this new direction and ended up on top. Jen has been on the cutting edge of spokesperson work, with clients like Microsoft. Not bad.

Fred Minnick is, quite simply, one of the nicest people I know. He also knows how to rock an ascot. Generous beyond measure, I was so lucky to have met him at an ASJA conference a while back. A kick-butt writer, his first book, *Camera Boy: An Army Journalist's War in Iraq* was a listed as a *Wall Street Journal* Best-Selling E-Book, and his second book, *Whiskey Women: The Untold Story of How Women Saved Bourbon, Scotch, and Irish Whiskey* is charming reviewers everywhere.

Marijke Vroomen Durning is a nurse turned writer. She's also a terrific presence when I'm feeling particularly cranky or sorry for myself. Her medical writing has shone an important light on the dangers of sepsis as she handles content management for the Sepsis Alliance. Take a look at her other work at TheNurseWriter.com. I share her obsession with fabric, though she does a better job of using it to make gorgeous quilts. By the way, here's how you say her Dutch name: muh-rye'-kah. (Bonus points if you can roll the *r*.)

Debbie Koenig and I published our first books at about the same time. Hers was *Parents Need to Eat Too: Nap-Friendly Recipes, One-Handed Meals, and Time-Saving Kitchen Tricks for New Parents*. She also generously offered me a guest spot on her terrific blog "Parents Need to Eat Too" writing about math. Debbie has been quite open to learning how everyday math is critical for writers and parents, and her enthusiasm is infectious.

Andrea Lynn knows what to do with some goat cheese and anything else you can find in the pantry. We cut our book-publicity teeth together, as our first books were published at about the same time. *The I Love Trader Joe's College Cookbook* is exactly what I needed, 25 years too late. It's hard not to admire Andrea's easy approach to cooking gourmet meals out of next to nothing.

Finally, **Sandra Hume** copyedited this book, which was no small task. Writing a math book is easy. Copyediting a math book is a real pain in the you-know-where. She carefully considered details like when to write numbers as words and when to use numerals. (That in and of itself was a monumental task.) And she worked through all of the math, bit by bit. Before this project, Sandra and I commiserated on such topics as raising extremely strong-willed girls. She's not only a terrific writer and editor, but also an amazing parent.

Of course I also must thank my family, Gina and Zoe. I appreciate your patience with my I'm-on-deadline crankiness, my stupid questions about math examples, and for putting up with my long hours and short temper. You both make me a much better person. I don't know what I'd do without you.

Introduction

IF YOU'RE READING THIS, chances are you have a degree in English or communications or journalism. You can spot a split infinitive from a mile away and may even edit your mom's emails without thinking. But math? Well, basic statistics and percentages may not be front-of-mind these days—if they ever were.

Fact is, math is an integral part of any job, even writing. From checking the numbers that a source gave to figuring out the best way to spend your book advance, you'll find it useful to know a number of simple calculations and a few more complex concepts. Sometimes the math separates the good writers from the *great* ones.

Since most writers have never taken a statistics course or haven't found a percentage change since middle school, you may need a little refresher. And if you don't have a strong math background, understanding how numbers work may feel intimidating.

But here's the good news: you are good at math. (Really!) Most people manage just fine with a basic understanding of percentages, formulas, and a dash of statistics. What has likely tripped you up is a lack of confidence, not an inability to understand mathematics.

And that's the point of this book. If you've ever felt overwhelmed by US Census data or the results of a big medical study, this book is for you. If you've ever wondered how you can jazz up your writing with some carefully—and creatively—placed numbers, keep reading. If you want to know how a few simple calculations can increase your chances of getting published, warm up your calculator. And if you want to increase your bottom line, get ready to prove your third-grade math teacher right—you will use math as a grownup. And you absolutely can do math.

Think of the hardest part of your job. Is it landing a great interview? Or trimming 5,000 words down to 1,200? Do you dread having to decide what happens to a character? Is a great lede sometimes elusive?

These are writer problems. We struggle over words and worry about accuracy. I'm not going to convince you that math will help you in these situations—at least not all of the time.

But what if you could focus on those things, rather than have an anxiety attack over whether you've calculated the percentage change correctly? What if you could greet a contract negotiation with enthusiasm, rather than worry that you're charging too little or too much? And what if you could quickly summarize a boatload of statistics with just one number—or better yet, a really cool analogy?

That would free your mind to focus on the reasons you became a writer in the first place—to write. Right?

This book will help you see how numbers can help your words come to life on the page (or screen). If you'd like to publish a book, you'll learn how to get a publisher to stand up and take notice with numbers that describe your platform and market research. If you're a freelancer, you'll become acquainted with easy ways to develop a project fee and track your financial goals.

Overcoming Math Anxiety

You may have had bad experiences learning about math or using math. If so, you may feel anxiety when you're called upon to work with numbers. Researchers have found that math anxiety affects girls and women more often than men, and it can set in at a very young age. If you're a woman with math anxiety, it's possible you got it from a female elementary school teacher—who was also math-anxious.

What are the symptoms? Well, in kids who are actively learning math every day, it looks like a lot of anxieties. They may get sweaty hands or feel jittery. Some may even throw up. Almost everyone describes a sense of their minds going blank.

When a kid—or a grownup—is so anxious that they can't remember what they've already learned, it's pretty darned hard to learn something new or perform on a test or calculate a tip at a restaurant. Right?

But here's the good news. If you have math anxiety, you can learn to manage it. First off, you've got to accept it. Be frustrated with it, if you want, but accept. This is not some sort of failure on your part. It just is.

Next up, face it, head on. New research shows that math-anxious students can reduce their symptoms just before taking a test by *writing about their feelings.* Cool, huh? If they spend a few moments scratching out a couple of sentences about their worries, they perform better on their tests.

You're a writer so why not try that neat trick next time you need to dive into a stack of company data or state test scores or interview a researcher? Take five minutes to simply write about your anxiety. Get it on paper and move on.

Finally, take these steps to really notice where you are *good* at math. Decide that for a day or two, you're going to keep general notes about when you use math. (This is a lot like keeping a food journal, for those of us who have been that route with adopting healthy eating habits.)

Don't cheat. Each and every instance of math—not just numbers, but mental and spatial calculations—gets noted. That includes figuring out if you have time to eat lunch before your next interview, or if you have enough butter in those 3 partially used sticks for a cookie recipe, or if you can scrape together enough nickels and dimes from your coin jar to pay the pizza delivery fellow.

Next, take a couple of moments to notice how you felt with each of these instances. Were you nervous? Did you notice you were doing math? If you didn't notice, pat yourself on the back. That's the goal here—to use math so effortlessly that you're not even aware of it.

Here's the thing: we all use math in very common ways—and very comfortably—because we *have to.* We become familiar with the math that makes our lives easier. The tough stuff is unfamiliar, and so we avoid it.

You can reach that easygoing place with more challenging mathematics—even percentages and statistics. Yes, you can. It takes some practice and maybe a little review. That's what this book is for.

Part 1 — Tell a Better Story

IF YOU'RE A REPORTER of any kind—whether freelancing for a parenting magazine or embedded with the 82nd Airborne Division in Afghanistan—you probably have two major goals: reporting the facts accurately and telling a good story.

Making sure the information is absolutely, 100% correct is paramount. Who cares if you have a good story if it's not true? But if no one's interested in reading the story, well, that's not worth much either.

Math can play a major role in both of these goals. You can check a source's numbers or evaluate a statistical study to decide if you can trust it. In fact, a little stats know-how can go a long way for writers reporting on health, business, education, politics, science, and more.

We've all read the "trend" stories that are less about trends than they are about anecdotal information uncovered by an insulated editor or writer. Or the shocking piece of data that turns out not to be true at all. These problems occur because the writer or editor didn't do the math.

Statistics is a branch of mathematics that can be quite peculiar and somewhat daunting, but having just enough understanding is important in most kinds of reporting.

And understanding statistics matters in other kinds of writing, too—from annual reports to self-help books to fiction. Like it or not, when writers avoid stats they can really screw up. This leads to all sorts of bad information and, worse, the reader losing trust.

So here's a cool secret that many writers don't know at all: some readers *look* for numbers to help them better understand a story. For them, it's not enough to say that a company's profit doubled. They want to know the starting figure and the ending figure. (And really, you should, too. If the starting profits are $0, doubling means nothing, nada, zip.)

In other words, you can really lasso your readers with a few carefully placed, well-thought-out figures. Don't shove them in there willy-nilly. Good writing includes numbers that help deepen the reader's understanding of your story.

And that's not all. With our ever-changing publishing industry, editors are looking for more from us writers. Knowing your way around some basic data displays—bar graphs, pie charts, or line graphs—can set you above the countless other writers pitching the big-name publications. Plus, many of us are self-publishing. In that situation, you—the writer—are in charge of charts and graphs.

Because again, readers want some math with carefully crafted metaphors, revealing quotes, and felt-like-I-was-there ledes. And *you* can give that to them.

Chapter 1 — Showing (Not Telling) with Numbers

MAKING SURE THAT THE story you're telling is accurate isn't the only reason to use numbers. Numbers offer ways to engage a reader's attention. But there are effective—and not so effective—ways to do this.

Think for a moment about how you read a news story, like this totally made-up one:

> *By the end of its first year, Jumpin' Jack's Pizza Joint had served 27,439 patrons, thanks to its premier location on The Boulevard.*
>
> *"We were slinging slices like there was no tomorrow," owner Jack Schloblanski says.*
>
> *In fact, he estimates that he made a total of 68,597 pies since opening his new location a year ago. At his old spot, he could count on making 23,000 pizzas each year.*

Be honest. Did you read it this way?

> By the end of its first year, Jumpin' Jack's Pizza Joint
> had served blah-blah patrons, thanks to its premier
> location on The Boulevard.
>
> "We were slinging slices like there was no
> tomorrow," owner Jack Schloblanski says.
>
> In fact, he estimates that he made a total of blah-
> blah pies since opening his new location a year ago.
> At his old spot, he could count on making blah-blah
> pizzas each year.

If so, you're not alone. Raw data doesn't always further the story. See, as writers, we're paid to clarify the information, not just dump numbers in just the way the source gives 'em to us.

Fact is, most people do not have a spot-on understanding of really large or really small numbers. You can count on your readers appreciating a few simple calculations that can help show the importance of the data—not just tell it straight out.

There are some marvelous ways to accomplish this while strengthening the writing and painting a more eloquent picture. Let's start with the first number.

Jack has served 27,439 customers in his first year at his new location. That's a lot of people—but what does it mean?

First, round the number so that it's easier to read and to manage. Numbers with zeros at the end or without a lot of digits on the right side of the decimal point are a lot easier to digest. So, if the actual number is not all that important, consider rounding. You can round 27,439 to 27,400, making the number easier for readers to read and process. Just be sure that you're not misleading the reader.

That brings us to another important point: when you round, make sure to let the reader know. Use words like "approximately," "about," and "a little more (or less) than."

Back to the pizza—after rounding the total number of customers served, you can express this value in a variety of different ways.

- Approximately 2,740 10-top tables
- About 75 customers every single day, including weekends and holidays
- Close to a 10-mile-long, single-file line

Really, the possibilities are endless. Depending on the style of writing, you can make numerical analogies that are funny, serious, or irreverent.

But how do you do the math?

Approximately 2,740 10-top tables

If you've ever worked in the restaurant business, you know that a 10-top is a table that seats 10 people. (A 2-top seats 2 people, a 4-top seats 4, and so on, and so on.) The math here is pretty darned simple: just divide 27,400 by 10 to get 2,740 10-top tables.

Want to take it a step further? Do a little digging to find out how many 10-tops would fit on a football field, for example. A quick Internet search shows that the area of a football field is 57,600 square feet. Catering companies suggest 15 square feet per person, in seated dining rooms with 10-top tables. That means 150 square feet per 10-top:

$$57,600 \div 150 = 384$$

So, 384 10-tops can probably fit in 1 football field. But 2,740 10-tops would require how many football fields? To find out, divide the number of 10-tops you need by the number that fit in 1 football field:

$$2{,}740 \div 384 \approx 7.14$$

What you've found is that if you seated everyone at 10-top tables, Jack's customers would take up a little more than 7 football fields!

You see how this can get really creative—and really out of hand.

About 75 customers every single day, including weekends and holidays

This little translation is a bit less creative, but pretty powerful. The math is super easy. Just divide the number of customers by the number of days in a year:

$$27{,}400 \div 365 \approx 75.07$$

A part of a person won't order pizza, so round down to 75.

While this is a really practical number, the writing can create even more of a picture.

> *On average 75 customers place an order at Jumpin' Jack's Pizza Joint every single day of the year.*

Want to make some allowances for holidays? Subtract them first and then divide. If Jack's is closed on Easter, Christmas, and New Year's, you'll divide by 362.

$$27{,}400 \div 362 \approx 75.69 \approx 76$$

(By the way, are you wondering what that wavy equals sign means? The $\approx$ indicates "approximately." In other words, $75.69 \approx 76$ means "75.97 is approximately equal to 76. It's important when you're showing that you've rounded. Remember it now?)

But this calculation can add a little detail that tells even more of the pizzeria's story.

On average 76 customers place an order at Jumpin' Jack's Pizza Joint every single day of the year—except for Easter, Christmas, and New Year's, the only days there's a CLOSED sign in the window.

What was originally a pretty easy way to break a big number down into a smaller one has now illustrated something important about the restaurant—it's closed only 3 days each year.

Close to a 10-mile-long, single-file line

For really visual readers, offering a vivid image is a great way to make sure that numbers hit their mark. Be sure, though, that what you're offering makes sense to the story.

In this case, getting in line for really good pizza is realistic. So why not think about how long the line would be if all of the customers queued up?

This is where a little creativity is necessary. Research by social scientists shows that our preferred personal space with friends is 1.5 to 4 feet. When around strangers, we may want much more than a 4-foot bubble. But standing in a single-file line means being a little more up-close-and-personal than that. So, it makes sense that most people are probably going to stand about 2 feet from each other. In other words, for every 1 person in line, the line will be 2 feet long.

Then I looked up the number of feet in a mile (5,280) and divided by 2. Why 2? Because I wanted to know how many 2-foot segments would fit into 1 mile.

$$5,280 \div 2 = 2,640$$

What this means, in essence, is that a single-file line of 2,640 people is 1 mile long. So how many miles is a single-file line of 27,400 people? Divide again.

$$27,400 \div 2,640 \approx 10.38$$

Yep, if you lined up 27,400 people, single file, giving each of them a 2-foot space to stand in, you'd have a line more than 10 miles long.

If Jumpin' Jack's has been dealing with long lines, this way of describing the numbers furthers the story:

> *For months, Jack has been managing long lines of customers waiting to be served or to place take-out orders. In fact if all of his customers lined up, single-file, by the end of the first year this line would be more than 10 miles long—or the distance from his current location to the new one.*

See what I've done there? The line metaphor works great, because it brings in another aspect to the story—that the new location is so busy, lines form outside regularly. For local readers, the metaphor is made even more meaningful by giving it some geographical context.

A word of warning: when you get creative, you run the risk of distorting the numbers—and misleading the readers. For example, if I decided that each person in a single-file line needed 5 feet of personal space, the imaginary line for pizza would be 25 miles long. But that's not really an accurate representation, is it? A little bit of research pointed me to a 2-foot bubble for each person in line—a much better estimation.

Images that illustrate numbers—like a 10-mile-long line for the pizza place—are really useful when numbers are huge and incomprehensible. For example, it's estimated that there are currently 200 billion pennies in circulation at this moment. That's a lot of pennies, but do you really know how many that is?

- $2 billion (or $2,000,000,000)

- 4.07 million cubic feet or enough to fill 98 floors (from floor to ceiling) of the Willis Tower in Chicago
- 625,110 tons or at least 1,905 flights on a Boeing 747 commercial plane (these planes can only carry 328.125 tons of weight)

There are *a lot* of pennies in circulation!

But although the just-right comparison can work well, the less-than-perfect comparison often falls flat and adds yet more confusion to what may be an already difficult story. As the writer, you should include the most effective analogies in your story. Test your idea on a friend or family member to see if it makes sense. Your editor's feedback may help you refine this illustration. Or you may need to push back, if your editor wants to change your comparison to something misleading or difficult to understand. If you go this route:

1. Make sure the image fits the story.
2. Make sure your readers can picture the references.
3. Avoid cliché images, like dollar bills placed end to end.
4. Don't add even more numbers with the image.
5. Don't repeat an image, like 23 football fields placed end to end. It's hard enough for a person to imagine the size of a football field. But to imagine 23 fields? That's asking too much of your readers.

Try it out. Some of your ideas will work, others won't. Get creative with the comparisons and then do the math.

Chapter 2 — Math Basics

NOW THAT YOU'VE GOTTEN your feet wet, adding a little math to your reporting and storytelling, it's probably a good idea to go back to the basics. If you're like most folks, it's been a long while since you rounded numbers, worked with negative numbers, or even found percentages. So here's a refresher.

Nice, round numbers

We've seen how rounding numbers can help you catch a reader's attention. Here's how it works.

1. Decide which place you want to round to, such as ones or tens. For example, if the number is 13,689, you might decide to round to the hundreds.
2. If the numeral to the right of that place is 5 or greater, round up by increasing the numeral in the place you want to round to by 1. Then set all the numerals to the right of that place to 0. For the above number, 13,689, that would mean making the "6" (in the hundred spot) into a "7" and turning the

"8" and "9" into zeros for a rounded number of 13,700.
3. If the numeral to the right of that place is less than 5, round down. Keep the digit you want to round to the same. Replace all of the digits to the right with zeros. So, if you have the number 13,215, and you want to round to the hundreds, you'll keep the "2" as is and reduce the "1" and the "5" to zero, for a result of 13,200.

Here's another example:

Round 341,979 to the nearest thousand

The thousand place is the digit directly to the left of the comma, which in this case is a 1. That's the number you'll be rounding up or down. Look to the right of the 1. There's a 9 there (which is 5 or greater, meaning you round up). Increase the 1 to a 2, and replace all of the digits to the right of 1 with zeros.

342,000

Same thing works with decimals:

Round 7.34092 to the nearest tenth

The tenths place is to the right of the decimal point. Look one more place to the right, and you have a 4. So you'll round down. 7.3000 is the same as 7.3, so for simplicity's sake, eliminate the unneeded zeros.

7.3

But how do you know when to round? If the numbers look like they'll get in the way of the story, round. For example, the US national debt might be $17,215,961,273,816.25, but that's an awfully confusing number. You don't sacrifice accuracy by reporting the national debt as $17.2 trillion. But if precision is important, keep 'em where they are. It makes more sense to report a state's minimum wage as $6.25 than to round down to $6.

Knowing what place to round to is important too. Consider the number $1,503,201,933. Should you round to the billion, making this $2 billion? Or should you round to the hundred million, making this $1.5 billion? There's a huge difference between $2 billion and $1.5 billion. A $500 million difference. For that reason, it's more responsible to round to $1.5 billion.

At the same time, even though our rule of thumb tells us to do otherwise, rounding 0.09945 to 0.1 is pretty darned close to 0.0995—and a heck of a lot easier to read.

Order matters

In a lot of places in this book, you'll notice that we're employing some simple mental math, but there's a lot going on behind the scenes. And one important step is the order that the calculations are done. This is called the *order of operations*.

In math, order matters—except when it doesn't. You need to know what those rules are and when you can break them. (Just like grammar.)

Remember Dear Aunt Sally? She was the key to the order of operations when most of us were kids: Please Excuse My Dear Aunt Sally or PEMDAS. What that translates to in mathematical terms is Parentheses, Exponents, Multiplication, Division, Addition, and Subtraction. That means you do the work in parentheses first, then calculate exponents, then do the multiplication, and so on. Here's how it works:

$$(1 + 4)^2 \bullet 6 - 4 \div 2 + 3$$

$5^2 \bullet 6 - 4 \div 2 + 3$	parentheses
$25 \bullet 6 - 4 \div 2 + 3$	exponents
$150 - 4 \div 2 + 3$	multiplication
$150 - 2 + 3$	division
$150 - 5$	addition
145	subtraction

Enforcing a particular order means that all of us get the same answer. If you calculated the operations in a different order—say, from left to right—you'd get something different:

$$(1 + 4)^2 \bullet 6 - 4 \div 2 + 3$$
$$(5)^2 \bullet 6 - 4 \div 2 + 3$$
$$25 \bullet 6 - 4 \div 2 + 3$$
$$150 - 4 \div 2 + 3$$
$$146 \div 2 + 3$$
$$73 + 3$$
$$76$$

Whoops!

A note about notation: see that little dot in there? If you remember way back to your school days, you know that $\bullet$ means multiplication. But why not use $\times$? I'm glad you asked.

First off, the x-shaped multiplication sign looks a lot like *x*, which is a very common variable. That's why your algebra teacher suddenly stopped using it, in favor of the dot.

But there's another reason that I think writers will appreciate. Typing math symbols is a giant pain in the *tuchas*. I can use keyboard shortcuts—option+8 on a Mac or Alt 0149 on a Windows machine—to make that dot, but there is no keyboard shortcut for $\times$. Instead, I've got to open up a special tool and insert that x-like symbol.

Now back to the order of operations: PEMDAS isn't some random set of rules. Mathematical theories prove why you need to multiply before adding, for example. But you don't need to know that to follow the order of operations.

But here's the rub. You can break some PEMDAS rules. We writers should get how this works. There are so many grammar rules we break—because we know we can. Sentence fragments. To boldly split infinitives in sentence fragments.

What you may remember from your early arithmetic lessons is this: multiplication and division are inverse operations. In other words, they undo each other. Addition and subtraction are, too. So the real truth is that you *don't have to* multiply before you divide or add before you subtract.

Another way to put this? PEMDAS = PEDMAS = PEMDSA = PEDMSA. Don't believe me? Try it for yourself.

Breaking PEMDAS Rules	Following PEMDAS Rules
$(1 + 3) \bullet 8 \div 2 + 9 - 1$	$(1 + 3) \bullet 8 \div 2 + 9 - 1$
$4 \bullet 8 \div 2 + 9 - 1$	$4 \bullet 8 \div 2 + 9 - 1$
$4 \bullet 4 + 9 - 1$	$32 \div 2 + 9 - 1$
$16 + 9 - 1$	$16 + 9 - 1$
$16 + 8$	$25 - 1$
24	24

Not sure when to break PEMDAS rules, still? Just follow Dear Aunt Sally's lead. When you get more practice, you'll become more comfortable with when and how you can switch things up—and still get the correct answer.

But isn't it nice to know that math is a little more flexible?

Comparing numbers: ratios, percentages, and proportions

Just like with metaphors and similes, a quick comparison can help some numbers make good sense. For example, I could tell you that 9 out of 10 writers recommend this book. But what does this mean? Well, 9 out of 10 can be translated several different ways. Here are some examples:

- 9:10 is a ratio
- $\dfrac{9}{10}$ is a fraction
- 90% is a percentage

Seeing 9 out of 10 in a variety of different ways gives you some flexibility in how you describe this value in your writing. For example, saying 9 out of 10 really emphasizes that of the 10 writers who were asked their opinion, 9 recommend this book. On the other hand, 90% implies that out of all of the writers in the world, 90% recommend this book.

The first two bullet points in the list above are *ratios*. In fact, in the second example, the ratio is written as a fraction. Of course 90% is something very familiar—a percentage.

Unless you're a sports writer—reporting lots and lots of game or match scores—you're going to spend a lot more time finding percentages than anything else. So let's review what these bad boys are.

Let's start by breaking down the word itself:

per - cent

Per means "each," while *cent* means "100." Ta-da! "Percent" merely means "per 100."

As you probably already remembered, 40% means 40 per 100. Let's say that 40% of the people who live in my city are cat owners. In other words, for every 100 city residents, 40 have at least 1 feline living in the house with them. You can state this comparison in a number of different ways: 40 out of 100 or 40 per 100. But finding percentages is much easier if you think of the word "per" as a division sign.

That is to say, 40% is the same thing as 40 ÷ 100. That fact is important because it means you can write a percentage as a decimal.

$$40 \div 100 = 0.40$$

Now, you don't need to plug 40 ÷ 100 into a calculator or scribble it on a piece of paper. Instead, what you're doing is simply moving the decimal point 2 places to the left.

Yes, the number 40 has a decimal point. But because 40 is a whole number, we don't generally write the decimal point. Just remember that it's directly to the right of the ones place—in 40, that means just to the right of the 0. So when you move the decimal point 2 places to the left, you move it to the left of the 4 to get 0.40.

A note about notation: when writing numbers with decimals, it can be a challenge to see that teeny-tiny dot. That means readers can misread ".4 percent" as "4 percent." To prevent confusion, you can add a zero to either side of a decimal point (as long as a digit isn't already there!). The zero doesn't change the value of the number, but it does make it easier to read. A reader is significantly less likely to mistake "0.4 percent" as "4 percent."

Once you've converted a percentage to a decimal, you can do other mathematical calculations with it. For example, if a source tells you that 15% of college students are African-American, and your research shows that 21.8

million people attended college last year, you can determine the actual number of students who are African-American. You would just multiply 21.8 million (total number of college students) by 0.15 (percentage of students who are African-American).

Just as a percentage can be written as a decimal, it can also be written as a fraction. Here's another way of representing 40%:

$$40\% \text{ is the same as } \frac{40}{100}$$

Later on, we'll explore times when expressing a percentage this way can help you make useful calculations.

Now that you know how to convert a percent into a decimal, you can find out all kinds of information. For example: what is 33% of 108?

$$0.33 \cdot 108 = 35.64$$

But what if you want to answer this question: 37 is 45% of what? What you are finding here is the whole—the larger value that 37 is 45% of. This can get a little confusing for some of us, but I've got a solution.

You can use 2 equal ratios—or a proportion—to solve *any* percentage problem. (Well, except for percentage change, which we'll get to later.) Here's how it works.

$$\frac{\text{percent}}{100} = \frac{\text{part}}{\text{whole}}$$

Take a close look at this proportion. There's a fraction on the left and a fraction on the right. The fractions are separated by an equals sign. That's a proportion.

Now look a little closer. Notice that the smaller bits— "percentage" and "part"—are in the numerators. (Numerator is just a fancy way to say, "top number of a fraction.") The bigger bits—"100" and "whole"—are in the denominators. (Denominator is just a fancy word for "bottom number of a fraction.")

Also notice that the values that relate to the percentage—"percentage" and "100"—are in one fraction, while the values relating to the actual data—"part" and "whole"—are in the other fraction.

In other words, there's a definite pattern here.

Noticing these details about the proportion means you don't have to memorize a formula or an algorithm (a specific set of steps to get to the right answer). It's much easier to look at a set of numbers and notice what they have in common, right? Then all you have to do is make sure that you've got "like" things in the same places. Make sense?

Don't worry if it doesn't yet. An example will help.

37 is 45% of what?

$$\frac{percent}{100} = \frac{part}{whole}$$

In the question, what is the percent? 45. What is the part? 37. And what about the whole? Well, that's what we're looking for. Substitute those values into the proportion.

$$\frac{45}{100} = \frac{37}{x}$$

Notice that there's an x in the "whole" space. That variable is there because we don't know what its value is. In fact, our goal is to solve for x or get it by itself.

Step 1: Cross-multiply

Multiply the numerator of the first fraction by the denominator of the second. That becomes the left side of the equation. Then multiply the denominator of the first fraction by the numerator of the second fraction. That becomes the right side of the equation.

$$\frac{45}{100} \diagdown \frac{37}{x}$$

$$45 \cdot x = 37 \cdot 100$$
$$45x = 3700$$

Step 2: Isolate x

To isolate x (get x by itself), you need to get rid of the 45. Since you're multiplying 45 by x, what do you have to do to get rid of the 45? Divide. Division is the inverse operation of multiplication. That means division "undoes" multiplication. In fact, $45x \div 45 = x$, which is exactly what you want!

But here's a really important thing to remember: whatever you do to one side of the equation, you have to do to the other. So if you divide the right side of the equation by 100, you have to divide the left side by 100, too. In this case, we're dividing both sides by 45:

$$45x = 3700$$
$$45x \div 45 = 3700 \div 45$$
$$x = 82.22$$

So, 37 is 45% of 82.22. Or stated another way, 45% of 82.22 is 37.

Now, are there other ways to accomplish the same thing? Yes! And if you'd rather go in a different direction, go for it. For example, you could simply multiply 37 by 100 and then divide by 45. (Do you see, that's what happened when solving the proportion anyway?)

But if you have trouble remembering what to multiply and then divide by, a proportion can be really useful. After all, it can be much easier to remember one proportion than to remember a variety of different algorithms or rules.

Pluses and Minuses

If you're like me, finding out about negative numbers rocked your 6th grade world. While negative numbers don't come up all that often in everyday grownup life, they do appear once in a while. Here's how to manage them, starting with multiplying and dividing—just because those are the easier operations.

Multiplying and Dividing Negative and Positive Numbers

1. If you multiply or divide 2 positive numbers, you get a positive answer
2. 2 negative numbers → positive answer
3. 1 positive number, 1 negative number → negative answer

If you make a calculation using 2 negative numbers, and you get a negative answer, then rule #2 tells you that something went wrong in the calculation.

These rules *only* apply to multiplying and dividing positive and negative numbers, so don't try this with addition and subtraction (or any other operations, like exponents). Some examples:

$$2 \bullet 4 = 8 \qquad 35 \div 7 = 5$$
$$-2 \bullet -4 = 8 \qquad -35 \div -7 = 5$$
$$-2 \bullet 4 = -8 \qquad -35 \div 7 = -5$$
$$2 \bullet -4 = -8 \qquad 35 \div -7 = -5$$

Math for Grownups

Adding Negative and Positive Numbers
1. If you add 2 positive numbers, you get a positive answer.
2. If you add 2 negative numbers, you get negative answer.
3. To add a positive and a negative number: ignore the signs and find the difference of the 2 numbers. (In other words, subtract the larger from the smaller.) The sign of the answer is the sign of the larger number.

Examples:

$$-12 + 7 = -5$$
$$12 + -7 = 5$$
$$-12 + -7 = -19$$

Subtracting Negatives and Positives
Subtracting is a lot like adding, but with a twist.

To subtract a positive and a negative number *or* 2 negative numbers: change the subtraction sign to an addition sign and change the sign of the second number to the opposite sign. (If the second number is positive, make it negative. If the second number is negative, make it positive.) Then add, using the addition rules above.

Examples:

$$23 - (-5) = 23 + 5 = 28$$
$$(-23) - 5 = (-23) + (-5) = -28$$
$$(-23) - (-5) = (-23) + 5 = -18$$

Chapter 3 — Percentages: A Writer's Best Math Friend

IF YOU LEARN NOTHING more from this book than how to find percentages, my writer friend, you are good to go. Percentages are the most important math tool many writers have.

Why? Because one way a writer can help readers get the message—and really understand a challenging, or not-so-challenging, topic—is to break it all down. This means comparing information in a way that readers can quickly grasp.

A comparison can help some numbers make good sense. This is where percentages come into play. And here's the really good news: working with percentages is way, way easier than you might think.

Percent vs. percentage

Let's get one little grammar detail out of the way. A *percent* is a number with a percent symbol (%). A *percentage* is the word we use to talk about these values when we're not using numbers.

In other words, there is zero mathematical difference between *percent* and *percentage*. No kidding!

Need an example? Here you go:

> *The company reported a 5 percent increase in revenue last fiscal year. The percentage increase was thanks to a booming economy and fortified sales staff.*

Is there a number? Use *percent*. Are you describing a value in a more general sense? Use *percentage*.

But wait! There's more! Whether or not you use *percent* or % is a style issue. As always, consult the style guide you are using for that advice. (For the curious: in this book, we're using %, except in passages that include news copy.)

Finding percentages and percentage change

Now let's put the math into action.

Meet Kayt, a crackerjack reporter covering the local dogcatcher election. Seems that the current officer of canine captivity suffered a small scandal when it was discovered he had 8 pups living in his 2-bedroom condo—4 more than the law allowed and 7 more than his landlord permitted. The city council called for a special election, and Kayt was brought in to report on it.

Like anywhere else in the US of A, voters have a tough time making it down to the local library or post office to vote in presidential elections, much less dogcatcher political contests. The local news station hired an independent company to conduct a poll, assessing whether residents were even aware of the upcoming election.

The results were not all that surprising to Kayt: 272 people said that they were aware of the special election, while 865 admitted to being clueless about the whole thing.

Kayt knows that if she only uses the raw data—the number of people who responded each way in the survey—she risks confusing her readers. The polling company didn't ask everyone, but the sample was a good one. That means the results can be generalized to the population of the whole town. (See the section on statistics for more info on that.)

So Kayt decides that converting this data into a percentage is her next step. What percentage of people knew about the special election, and what percentage did not?

To find out, Kayt needs to know the total number of people who responded to this poll. Pretty simple: 272 + 865 = 1,137.

With that value, Kayt can calculate the percentages by dividing the *part* (those who knew about the election *or* those who didn't know about the election) by the *whole* (everyone who was polled).

Respondents who did know about the dogcatcher special election:

$$272 \div 1,137 = \frac{272}{1,137} = 0.239 \approx 24\%$$

(Psst—there's a very good reason that Kayt's written 272 ÷ 1,137 as a fraction. As I mentioned earlier, it means the very same as using a division sign [272 ÷ 1,137], and it helps her keep track with more complicated percentages. Stay tuned.)

Respondents who weren't aware of the special election:

$$865 \div 1,137 = \frac{865}{1,137} = 0.761 \approx 76\%$$

Notice that Kayt has decided to round her percentages. In this case, she decides that rounding works just fine for the story. But she does want to check her results. If she adds both percentages, she should get 100: 24% + 76% = 100%. Kayt's good to go.

And now she's got some more revealing numbers. More than three-quarters (76%) of the people polled don't have a clue about the special election.

In an interview with a city official, Kayt learns that the city council is concerned enough about the special election that they're sponsoring a get-out-the-vote campaign. Through mailers and a few carefully placed television and radio ads, each of the candidates will get a little nonpartisan, free advertising.

Interesting, Kayt thinks. And she decides to revisit this story after the ads run and brochures land in mailboxes. Because this new poll uses the same questions and surveys the same number of people, Kayt can compare the results.

Two weeks later, the results are in—but in a different format.

- 62% of those polled now say they are aware of the campaign, and
- 38% are still in the dark.

No raw data this time. Still, Kayt would like to consider the percentage change from the old poll to the new poll. In order to do this, she needs to know how many people are aware of the campaign and how many are not.

She could go back to the group conducting the poll, or she could employ a little bit of math. Kayt pulls out a blank piece of paper and gets going.

Here's where writing the percentage as a fraction helps a lot. In fact, as I pointed out in Chapter 2, the easiest way to solve this problem is to use a proportion (two equal fractions or two equal ratios).

Remember how I mentioned before that Kayt needed to calculate percentages by comparing the part to the whole? This is how that statement can be expressed mathematically:

$$\frac{\text{percentage}}{100} = \frac{\text{part}}{\text{whole}}$$

Kayt knows the *percentage* of respondents who know about the election, the percentage who don't know about the election, and the total number of respondents. What does she want to know? The *number* of respondents who answered a particular way.

She can create a proportion to address her questions. For example, she wants to know the number of people who are aware of the election (the part of the whole). She knows the percentage (62). She plugs that in on the percentage side of the proportion. She knows that the denominator for a percentage is always 100 (a percentage is always some part of 100), so she writes that in. She also knows the total number of respondents, so she plugs that number into the part/whole fraction side of the proportion. Since the total number of respondents equals the whole, she puts that in the denominator. That leaves the part (number of people who know about the election) as unknown. She indicates that with a handy-dandy *x*:

$$\frac{62}{100} = \frac{x}{1,137}$$

Remember, that unknown is called a variable. You don't have to call it *x*. You can call that whatever you want—draw a smiley face, call it *Fred*—but it might make things a little easier using a variable like *x*.

Now for the algebra. You have to get *x* by itself on one side of the equal sign (then you'll have a number that equals *x*, thus solving your problem). To isolate ("find") *x*, start by cross-multiplying.

62% of respondents know about the election. How many people is that?

$$\frac{62}{100} = \frac{x}{1,137}$$

Step 1: Cross-Multiply

$$62 \cdot 1,137 = 100 \cdot x$$
$$70,494 = 100x$$

Step 2: Isolate x

$$70,494 \div 100 = 100x \div 100$$
$$704.94 = x$$

Clearly 704.94 people cannot respond to a telephone survey, so Kayt rounds this value to 705. That means 705 of the 1,137 respondents now say that they have heard of the upcoming dogcatcher election.

Kayt repeats the process to find out the number of people who still don't know about the election.

38% of respondents report that they don't know about the special election. How many people is that?

$$\frac{38}{100} = \frac{x}{1,137}$$
$$38 \cdot 1,137 = 100 \cdot x$$
$$43,206 = 100x$$
$$43,206 \div 100 = 100x \div 100$$
$$432.06 = x$$

According to Kayt's calculations, about 432 people still do not know about the special election. Again, she checks his numbers, making sure that these numbers add up to the total people surveyed: 705 + 432 = 1,137.

You may be looking at Kayt's calculations and thinking: why didn't she just find 0.38 • 1,137? Well, she could have. And if that's how you remember how to solve these kinds of problems, go for it.

But if you have a hard time remembering when to multiply or divide when solving a percentage problem—or what to multiply or divide by—this proportion can be really handy. In fact, you can solve *any* percentage problem using a proportion, making it a really powerful tool.

To organize all of the information that she's gleaned so far, Kayt decides it's time to put everything into a nice, neat table.

(In fact, that's a great tool that mathies use. Organizing and labeling what you know can help you keep track of everything. And tables are great for this.)

Responses	Poll 1, raw data	Poll 1, percentages	Poll 2, raw data	Poll 2, percentages
Know about the special election	272	24%	705	62%
Don't know about the special election	865	76%	432	38%
Totals	1,137	100%	1,137	100%

Clearly, after the city council's awareness campaign, the number of people who recognized at least one of the candidates for dogcatcher increased. But by how much?

Kayt needs to determine the percentage change—increase or decrease—of the respondents who know about the election and those who don't know about the election.

Percentage change is the percentage by which a value increased or decreased. These figures are especially useful

29

when examining values over time. If a company increases its profits from last year to this, the percentage change would show by what percent profits grew. If a person's body fat percentage decreased from one month to the next, the percentage change would by what percentage his body fat decreased.

The percentage change is based on the difference between the old value and the new value, BUT (and this is important!) you can't just subtract the old percentage from the new one to get your answer. Nope, that's not the percentage change, that's the percentage *point* change (something entirely different, which we'll discuss in a bit).

In fact, when you calculate percentage change, you don't use percentages at all. You actually need just two values: the old value and the new value. These are the raw numbers, not percentages.

Here's how it works, using simple language:

(new value – old value) ÷ old value

In other words, subtract the old value (for example, the *number* of people who didn't know about the special election—you don't use the percentages in this calculation) from the new value (the number of people who still don't know about the special election despite the awareness campaign), and then divide by the old value.

The old value is the number before the change. In this case, the old value comes from the first poll taken.

The new value is the number *after* the change. In this case, the new value comes from the second poll taken.

"But wait!" you might say. How can you tell if you're finding a percentage increase or a percentage decrease? Watch.

Kayt first considers the number of poll respondents who were *not* aware of the dogcatcher race. In the first poll, that was 865 people. In the second poll, it was 432 people. She uses a little common sense first: since the number of respondents in this category decreased, she's looking for a percentage decrease. Her answer should show that clearly. And it does. Check it out:

$$(\text{new value} - \text{old value}) \div \text{old value}$$
$$(432 - 865) \div 865$$
$$-433 \div 865$$
$$-0.501 \approx -50\%$$

The answer is negative, which confirms what Kayt already knew—the percentage change is a percentage decrease.

What about the percentage change in respondents who were aware of the election? It should be an increase, since in the first poll this number was 272 and in the second poll it was 705.

$$(\text{new value} - \text{old value}) \div \text{old value}$$
$$(705 - 272) \div 272$$
$$433 \div 272$$
$$1.592 \approx 159\%$$

The answer is positive, so yep, there's a percentage increase in the number of people who now know about the election. In fact, Kayt recognizes that because the percentage is greater than 100, the campaign more than doubled the awareness of the election. Not bad.

Further, because the percentage is greater than 100, Kayt might consider writing it differently. For example, 150% is the same as 1.5 times as many, and 200% is the same as twice as many. Kayt could say that awareness of the campaign grew by about 1.5 times, since 159% is close to 150%.

Kayt glances back up at his table and sees that just with the raw data, she can tell that her calculations are correct.

705 is more than twice 272, and 432 is a bit more than half of 832.

Of course none of this information shines light on whether people will actually vote in the special election— or whom they'll vote for. And it doesn't show that the advertising campaign *caused* greater awareness. But it does say one thing for sure: there's a correlation between telling people about an election and actually getting out the vote. (More on the difference between correlation and causation in the chapter about statistics.)

A few good rules

When working with percentages, keep the following facts in mind:

1. Percent means *per 100*. So to convert a percentage to a decimal, just divide by 100. This is the same thing as moving the decimal point 2 places to the left. For example: $35\% = 35 \div 100 = 0.35$

2. To find what percentage a number is of the whole, divide the part by the whole. A good way to think of this is as a fraction:

$$\frac{\text{part}}{\text{whole}}$$

3. When you know the percentage but are being asked to find the part or whole, a proportion can be really helpful:

$$\frac{\text{percentage}}{100} = \frac{\text{part}}{\text{whole}}$$

To solve this proportion, substitute the values that you know. Then cross-multiply and solve for the unknown. Example:

What is 63% of 1,467?

$$\frac{63}{100} = \frac{x}{1,467}$$

$$63 \cdot 1,467 = 100 \cdot x$$

$$92,421 = 100x$$

$$924.21 = x$$

924.21 is 63% of 1,467

4. To find the percentage change, subtract the old value from the new value and divide by the old value, like this:

(new value – old value) ÷ old value

Example: you sold 5,387 books last year and 6,923 books this year. What is the percentage change of book sales from last year to this year?

$$(6,923 - 5,387) \div 5,387$$

$$1,536 \div 5,387$$

$$0.285 \approx 29\%$$

Interpreting percentages

Often sources give percentages to journalists. It's our job to do the interpretation, and unfortunately a lot can go wrong.

Percentage points

When reporting on the banking industry, an intern wrote:

Average interest rates on personal savings accounts have grown by 1 percent since last year, from 2 percent to 3 percent.

But this is wrong. The interest rates haven't increased by 1%. They've increased by 1 *percentage point*.

When you subtract numbers expressed as percentages, the result is a percentage point difference, not a percentage change.

In the example above, the interest rates have increased by 50%. Here's why.

The interest rates started at 2%, right? And then they grew to 3%. So how many percentage points did they increase by? That math is incredibly simple:

$$3 - 2 = 1$$

Got it? But that's not 1%. It's 1 percentage point.

Now, let's compare the original percentage—2%—and that increase—1 percentage point. Do you notice anything? Yep, 1 is half of 2. So the interest rates increased by one-half or 50%.

Let's look at another example. If the president's approval rating has dropped from 30% to 28%, it's fallen by 2 percentage points. But what is the percent change of his approval rating?

To find out, subtract the old approval rating from the new approval rating and divide by the old approval rating:

$$(0.28 - 0.30) \div 0.30$$

$$(-0.02) \div 0.30$$

$$-0.066667 \approx -7\%$$

So even though the president's approval rating has only dropped by 2 percentage points, it's actually fallen by 7%.

But there are times when it's a good idea describe the percentage point change rather than (or in addition to) the percentage change. For example: "Interest rates rose from 10% to 12%, a 2 percentage point increase that translates to a 20% increase in interest payments."

Teeny-tiny percentages

The World Health Organization estimates that cystic fibrosis is appears in 1 of every 3,500 births in the US. As a percentage, that's:

$$\frac{1}{3,500} = 0.00028571 \text{ or } 0.028571\%$$

That's a teeny-tiny percentage, less than 0.03%. It may be worth it to leave it at 1 of every 3,500, rather than converting to a percentage. At the same time, parents who are worried about this particular disease may feel more assured knowing that their chances are less than 0.03%.

The point is this: consider carefully how you will describe really small percentages. Offer readers more than one way to understand the numbers. You never know how the information may be best digested. So it's a great idea to offer options.

Apples to apples

Whenever you're comparing percentages, make sure you're comparing "like" things. Take a look at this example.

Ten percent of Republicans voted for the new beverage tax, while only 8 percent of those who voted in favor of the beverage tax were Democrats.

This one is a little trickier to see. Take a look at the first part of the sentence.

Ten percent of Republicans voted for the new beverage tax.

Now the second half of the sentence.

> *Eight percent of those who voted in favor of the beverage tax were Democrats.*

Look for the word "of" in each of these. What's the phrase behind it? Think about the words used to describe percentages. "Of" means "divide," so whatever comes after "of" is the number on the bottom of the ratio.

$$\frac{10}{100} = \frac{x}{\text{Republicans who voted for the tax}} \text{ and}$$

$$\frac{8}{100} = \frac{x}{\text{those who voted for the tax}}$$

With these two proportions, the bottom values of the second and third ratios are not the same. And if they're not the same, you can't compare them.

In order to make an accurate comparison, you need to have one of the following values:

1. The number of Democrats who voted for the tax, *or*
2. The number of people who voted for the tax who are Republican.

In this case, you've got to go back to the data or go back to the source for more data.

Projecting changes

Another danger zone is projecting changes. Let's say that a state's deficit is expected to grow by 15% each year, if spending and taxation remain the same. In 10 years, by how much will it have grown?

You can't just multiply 15 by 10 to get the answer. The deficit will *not* grow by 150% after a decade.

Here's why, using easy numbers. Assume the state deficit is $100 at the beginning of year 1. At the end of that year, the deficit has grown by 15%, so the total deficit is

$115. At the end of year 2, the deficit will have grown another 15%. But 15% of $115 is not $15 (as it was in year 1). In year 2, the deficit is actually $17.25. That's because 15% of $115 is $17.25: 0.15 • 115 = 17.25.

What you're doing here is compounding the change, which is exactly how compound interest is found. Each time you find the percentage, you add the result to the previous amount. The next time you find the percentage, you're finding it on the new—larger—amount. You could do this, over and over, until you get to the year that you want, or you could use a formula.

$$\left(\left(1+r\right)^t -1\right)\bullet 100$$

Yes, this thing is uh-gly. But if you break it down into smaller pieces and remember the order of operations (Dear Aunt Sally), it's not too bad. Promise.

(Also, there's no need to memorize this formula. If you do this kind of calculation often, just jot it down on a Post-it® and put it on your desk. Memorizing formulas is for middle-school students. Your brain is used for more important tasks now.)

First, let's identify the variables. The r stands for *rate* or the percentage. In this case, we're talking about 15% or 0.15.

The t stands for *time*. The interest rate is increasing annually, and you want to know what happens after 10 years. So t = 10.

Substitute, and then we can do some simple arithmetic.

$$\left(\left(1+0.15\right)^{10} -1\right)\bullet 100$$

Think back to Dear Aunt Sally (or whatever you use to remember the order of operations). Then follow the steps. You need to deal with the parentheses first, but which ones? Start with the ones on the very inside.

$$\left(1.15^{10}-1\right)\bullet 100$$

Now deal with the stuff in the remaining parentheses—an exponent and subtraction. The order of operations says to calculate exponents before subtracting.

$$\left(4.0146-1\right)\bullet 100$$

$$3.046\bullet 100$$

The last bit is really easy. Just multiply by 100 to get 304.6%. So at a growth of 15% per year, the state's deficit is expected to grow by three-fold after 10 years.

Big difference, eh? That's because compounding makes the total grow much more quickly.

And with that short review, you know enough about working with percentages to use them to help you tell your story.

Chapter 4 — Lies, Damned Lies, and Statistics

MARK TWAIN HAD IT right. Statistics can certainly pervert a message, either intentionally or unintentionally. And then there's the whole fact that stats are a highly peculiar arm of mathematics that can be quite confusing and difficult to comprehend.

But out of all of the math in the world, statistics is the most important for writers to understand—or at least know they need help with.

So what is statistics? In its most basic form, statistics is the practice of collecting and analyzing numerical data in large quantities.

There are two broad categories of statistical procedures: descriptive and inferential statistics.

See, some studies and polls are designed only to find out *how many*—how many people are planning to vote, how many birds are killed by cats each year, how many free or reduced lunches are served in a school system.

This is called *descriptive statistics.* The numbers may be presented as a percentage or an average. They may be illustrated with a graph or an infographic.

But notice that only the data is represented. No one is connecting A to B or showing causation. That's the job of *inferential statistics*, a much more complicated process, because probability is thrown into the mix.

It is really not important for most writers to grasp the details of inferential statistics. But there's one unbelievably important point to make here. Making inferences is not about a gut reaction or what looks like common sense. Instead, this is much more like the scientific process.

In inferential statistics, a hypothesis is often developed and then tested, using a variety of statistical tools, like *p-value*, *confidence interval*, and *Bayes factors*.

(Hey! Snap out of it! I saw your eye glaze over. You don't need to know any of those inferential statistics thingies. Leave that to the experts. It's enough to know that you can't draw reliable conclusions *without* them. Okay? Feeling better?)

Here's the Most Important Thing about statistics, so listen good. Descriptive statistics can suggest a correlation. Data can show that two things are related. But that's where it stops. Only inferential statistics can suggest causation.

Each of these parts requires very specific sets of skills and tools. A particular conclusion might make common sense, but that doesn't mean it's statistically sound.

In 1999, *Nature* published the results of a University of Pennsylvania Medical Center study of nearsighted children. The research seemed to show that sleeping with the light on as an infant caused myopia. This conclusion turned out to be false. While there is a link between a parent's and biological child's vision, sleeping with the light on does not cause nearsightedness. In fact, it could very well be that nearsighted parents leave the nursery light on, so that they can better see when they're awakened for midnight feedings or diaper changes.

It's pretty easy to misunderstand even the best studies. But when statistics is used properly, it is, in and of itself, a way to tell a story.

So we know that a lot of women are now surviving breast cancer. In fact, current research shows that 89.5% of all women diagnosed with breast cancer live for at least 5 years after diagnosis. But how many were originally diagnosed? At what stage were they diagnosed? What kinds of treatments did they get? What kind of preventative care did they have? And what do these things mean for the rest of us?

If I write a story about breast cancer that only describes survival rates, I'm not telling the whole story. And maybe that's okay. But unless the research says it's so, I should not connect better diagnostic tools or treatment methods to greater survival rates. In other words, there may be a correlation but no causation. In fact, there could be a thousand other possible reasons—some that scientists haven't even considered yet—for the increase in survival rates.

The processes that researchers go through in order to get reliable information that leads to reliable conclusions are driven by mathematics and science. Getting good at this process takes years and years of education and practice.

The good news is that you don't have to go through all of that. Aside from a little bit of arithmetic, you don't even have to do any calculations in order to use statistics well in your writing.

Why care about stats? First off, you do your readers a big favor by giving number-based details. But there's more to it than that. These days, our society is swimming in data. Big Data.

Big Data is what the grocery store is collecting when you use your membership card before you check out—what you purchased, how much you paid, what coupons you used, what time you shopped, etc. Then this data is tied to information about you, like your age, gender, and zip code. Big Data also comes from social media platforms, like Facebook, Twitter, and Google+, including what you "like," keywords in your status updates, tweets, and posts, when you log on, etc. Big Data is how Target outed a pregnant teen to her father by mailing her a flier advertising discounts on cribs and maternity wear. (This is a true story. While it's unclear what Target's algorithm was exactly, it seems that the chain store assigns women a "pregnancy probability score" based on their age, purchase history, and information purchased from other data miners.)

Thanks to Big Data, we are soon to be inundated with much more information about the world we live in. Because of this, we writers will have much more information at our fingertips. We can use that data to bolster arguments in our stories, push sources to be more clear, and even (as you'll find out in later sections) get published.

But having lots and lots of data doesn't mean that it's used well by writers, editors, and publishers—or even by sources themselves. Too often the information is put into misleading graphs or interpreted in a way not intended by researchers. Sometimes this is unintentional, and sometimes it's done on purpose.

Which is why you need some stats skills—to spot the fakes or crunch the numbers yourself.

Statistics, you're so random

Look, you can't poll every single person in Utah to find out what their thoughts are on the next presidential race. A university can't track every infant in the world who sleeps with a light on, to see which ones become nearsighted. And that's precisely the point of statistics. They can help us reach conclusions even if we can't count everyone.

First, some definitions. A *population* is the entire group that the researchers are interested in. So, if a school system wants to know parents' attitudes about school starting times, the population would be all parents and caregivers with children who attend public school in that district.

A *sample* is a subset of the population. It would be nice to track the viewing habits of all television viewers, but that's just not a realistic endeavor. So A.C. Nielsen Co. puts its set-top boxes in a sample of homes. The trick is to be sure that this sample is big enough and that it's representative.

A *member* is one individual (person or thing) in a population or sample. It's tempting to think of "members" as people, but statistically speaking a member could be anything at all that is being represented by the population or sample. If the population is all grizzly bears in Wyoming, a member is one of those grizzly bears. If the sample includes every fifth car that comes off an assembly line, a member is one of those cars.

Turns out, a large enough, random sample is representative of the whole population. Pretty cool, eh? In fact a good sample is at least 1,000 members and is completely and utterly random. That random thing is tougher than it looks, which means it is *really* easy to collect bad statistics.

This is something that every single special interest group knows well. Some are just fine with misleading the public in order to get what they want. Others are much more responsible. And still others just make mistakes.

As a writer, it's your job to recognize the bad stuff and avoid it like the plague. The sample is a great place to start.

Size Matters

If the entire population is small enough—say students in a classroom or members of a book club or new saplings at a tree farm—sampling is not necessary. When all members of a population are included in the survey or poll, what you have is a census.

But even the US Census isn't a real census. While the idea is for 100% participation, that's a tall order for a population of about 319 million people. In that case, sampling the population is the best bet.

Sampling is important in a variety of different settings. A Complete Blood Count (CBC) requires a sample of your blood—not all of it. A medical trial features a sample of people, not everyone in the population. A political pollster only contacts a sample of people for their opinions, not everyone in the district, state, or country.

So what is a good sample size? One million? Ten thousand? Five hundred? Seventy-five? When there are individual members who can be counted, the number is actually not that large. As long as the sample is random, 1,000 members is a good number to shoot for. And if you're talking about a local poll, 350 to 450 respondents are enough.

Surprised? Statisticians have looked at this process 17 ways to Sunday, and the results are pretty darned consistent. Larger samples don't get you any closer to a true representation of the population.

And this is great news for the outfits that conduct this research. It's a lot cheaper to poll 1,000 people than to poll 10,000 people. It's a lot less work to get 1,000 people in a medical trial than 100,000 people.

But researchers might deliberately choose smaller samples. Preliminary medical trials are often much smaller than 1,000 people. Instead, they may test as few as 50 people. This can offer a great deal of insight for the researchers. However, it's critical to note that the results cannot be generalized to the rest of the population.

And some pollsters simply ignore sample size, either out of ignorance or to purposely mislead the public.

When reading a study or poll, check the sample size. If it's too small, you, dear writer, must resist making the assumption that the results do apply to the greater population.

Getting Random

When samples don't represent the larger population, the results aren't worth a darn. And that's where randomness comes in.

So what does it mean for a selection to be random? Think of flipping a coin or choosing a name from a hat or even pulling a lottery ball. Those are all random events.

A random or independent event is one that is not affected by other events. Take a coin flip, for example. Each time you flip a coin, you have the same chance of getting heads as you do tails. (We're assuming here that you have a fair coin. In other words, the coin is not weighted so that one side or the other shows up more often.) If you toss the coin 15 times, each time getting heads, with the next toss you still have a 50-50 chance of getting heads.

Lots of you are thinking, "Wait! But if I get 15 heads in a row, aren't I due a tails?" Nope. That's called the Gambler's Fallacy—and it's what keeps the lights on in Vegas.

Whether you're spinning the roulette wheel or pulling a slot machine arm or rolling the dice at a craps table, each event is independent. This is precisely what makes them random.

In order to be sure that the results are reliable, researchers must select their samples using a random process. But first, they have to be sure that the entire population is represented. That's the really hard part.

Ever hear of President Landon? Nope? There's good reason for that. But on Halloween 1936, a *Literary Digest* poll predicted that Gov. Alfred Landon of Kansas would defeat President Franklin Delano Roosevelt come November.

And why not? The organization had come to this conclusion based on an enormous sample, mailing out 10 million sample ballots, asking recipients how they planned to vote. In fact, about 1 in 4 Americans had been asked to participate, with stunning results: the magazine predicted that Landon would win 57.1% of the popular vote and an Electoral College margin of 370 to 161. The problem? This list was created using registers of telephone numbers, club membership rosters, and magazine subscription lists.

Remember, this was 1936, the height of the Great Depression and long before telephones and magazine subscriptions became common fixtures in most families. *Literary Digest* had sampled largely middle- and upper-class voters, which is not at all representative of the larger population.

On Election Day, the American public delivered a scorching defeat to Gov. Landon, who won Electoral College votes in Vermont and Maine only. This was also the death knell for *Literary Digest*, which folded a few years later.

When the sample is not representative, the results can be biased. In fact, this example neatly describes two forms of sample bias: *selection bias* and *nonresponse bias*.

Selection bias occurs when there is a flaw in the sample selection process. In order for a statistic to be trustworthy, the sample must be representative of the entire population. For example, conducting a survey of homeowners in one neighborhood will not be representative of all homeowners in a city.

Self-selection can also play a role in selection bias. If a poll, survey, or study depends solely on participants volunteering on their own, the sample will not necessarily be representative of the entire population. There's a certain amount of self-selection in any survey, poll, or study. But there are ways to minimize the effects of this problem.

Nonresponse bias is related to self-selection. It occurs when people choose not to respond, often because doing so is too difficult. For this reason, mailed surveys are not the best option. In-person polling has the least risk of nonresponse bias, while telephone surveys carry a slightly higher risk.

If you're familiar with information technology, you know the old adage: garbage in, garbage out. This definitely holds true for statistics. When the sample is bad, the results will be, too, but that doesn't stop some from unintentionally or intentionally misleading the public with bad stats. If you plan to make good decisions at any point in your everyday life, well, you'd better be able to cull the lies from the good samples.

So what can you as a writer do to avoid reporting data from unrepresentative samples? First off, choose your sources wisely. Some reputable pollsters include Gallup, Roper, Harris Interactive, and Mason-Dixon, along with most big media and university polls and surveys.

Look at the source of the poll. Any polls conducted by a candidate, special interest group, or company with a stake in the outcome should be avoided. Ask who sponsored or paid for the poll. If they won't tell you, don't use it.

Reliable research comes from organizations like the US Census, National Institute of Health (NIH), Centers for Disease Control (CDC), Government Accountability Office (GAO), and various reputable nonprofit health organizations and teaching hospitals.

Track down the study or poll whenever possible. In other words, don't assume that because it was reported in a reputable source, the information is correct. It makes no difference if *Newsweek* or *The Washington Post* or *The New York Times* wrote it first. The worst mistake is the one that's repeated over and over.

A sampling of samples

There are several kinds of samples. The type that is chosen determines the reliability of the data. The more random the selection of samples, the more reliable the results. Here's a rundown of several different types:

Simple Random Sample: The most reliable option, the simple random sample works well because each member of the population has the same chance of being selected. There are several different ways to select the sample—from a *lottery* to a *number table* to *computer-generated values*. (Knowing what these are isn't all that critical for writers, but if you're curious, I explain what a lottery, number table, and computer-generated values are on my blog. Visit www.mathforgrownups.com and search "simple random sample.")

After a member has been selected for the sample, one of two things can happen. It can be put back into the population, for a chance to be selected again. Or it can be held out, so that there are no duplicate selections.

Stratified Sample: In some cases it makes sense to divide the population into subgroups and then conduct a random sample of each subgroup. This method helps researchers highlight a particular subgroup in a sample. This process is particularly useful when observing the relationship between two or more subgroups. However, it's important that the number of members selected from each subgroup match that subgroup's representation in the larger population.

What the heck does that mean? Let's say a researcher is studying glaucoma progression and eye color. If 25% of the population has blue eyes, 25% of the sample must also. If 40% of the population has brown eyes, so must 40% of the sample. Otherwise, the conclusions may be unreliable, because the samples do not reflect the entire population.

Then there are the samples that may not provide such reliable results:

Quota Sample: In this scenario, the researcher deliberately sets a quota for a certain group. When done honestly, this allows for representation of minority groups of the population. But it also means that the sample is no longer random.

For example, if you wanted to know how elementary-school teachers feel about a new dress code developed by the school district, a random sample may not include any male teachers, because there are so few of them. However, requiring that a certain number of male teachers be included in the sample ensures that male teachers are represented—even though the sample is no longer random.*Purposeful Sample*: When it's difficult to identify members of a population, researchers may include any member who is available. And when those already selected for the sample recommend other members, this is called a *Snowball Sample*.

While this type of sample is not random, it is a way to look at more invisible issues, including sexual assault and illness.

Convenience Sample: When you're looking for quick and dirty, a convenience sample is it. Remember when survey companies stalked folks at the mall? That's a convenience or accidental sample. These depend on someone being at the right (wrong?) place at the right (wrong?) time. When people volunteer for a sample, that's also a convenience sample.

So whenever you're looking at data, consider how the sample was formed. If the results look funny, it could be because the sample was off.

Correlation and Causation

Remember those nearsighted kids who slept with the light on when they were babies? As I mentioned earlier in this chapter, just because two events are correlated doesn't mean that one causes the other.

It is so tempting to look at data and jump to conclusions. But remember that causation is established (or refuted) using inferential statistics, a set of steps that ensures the conclusion is valid and not just coincidental.

To explore this further, let's look at an example.

VROOMS—Very Rad Organization of Motorcycle Safety—conducts research on motorcycle helmet use in several different regions of the country. After crashing his bike several times in the desert, Fred is a convert. Helmets are a must, and he's volunteered with VROOMS as a data collector. Hundreds of volunteers like Fred sign up throughout the country, to gather information about helmet use.

The process is simple. Fred is stationed at a busy road, where he counts each motorcyclist who passes. At the same time he makes one of three marks: H, I, NH. When the biker is wearing a helmet correctly, he writes down an H. When the biker is wearing a helmet but there's a problem, like a bad fit or the strap isn't fastened, an I goes in the little box. No helmet? That's an NH.

Fred stands at his post for eight hours, checking off motorcyclists on his clipboard. He carefully tabulates the data and then sends the results, plus the check-off charts, to VROOMS' main office.

On that one day, Fred saw 48 motorcyclists. Seven of them were not wearing helmets (NH); 23 were wearing helmets, but there was some sort of problem (I);and the rest were wearing helmets properly (H).

"Whew," Fred thought. "There's a big problem with helmets in my area." Converting to percentages helps show this clearly.

$$\text{NH: } 7 \div 48 = 0.146... \approx 0.15$$

$$\text{I: } 23 \div 48 = 0.479... \approx 0.48$$

$$\text{H: } 18 \div 48 = 0.375 \approx 0.38$$

(Quick check: where did the 18 come from in the last percentage? That's the number of motorcyclists who were wearing helmets. 7 + 23 = 30 and 48 – 30 = 18.)

So about 15% of the motorcyclists that Fred saw that day were not wearing a helmet, about 48% were wearing one incorrectly, and about 38% were wearing a helmet properly.

(Yes, this adds up to 101%, which is too much. Do you see why, though? It's the rounding. So it's not a big deal.)

Fred is right, there's a big problem with helmet safety among motorcyclists. VROOMS notices similar results in other parts of the country, so it launches an awareness program. In six months, the group will send out volunteers to see if there is a change in the data collection.

So there Fred is again, sitting in a lawn chair by the side of the road, with his trusty clipboard and pen in hand. It's been 6 months since VROOMS began airing radio commercials, bought advertising in local newspapers, and sent educators to local motorcycle groups and shops. He's fully expecting to see a huge change in his results.

Sure enough, he does. At the end of his eight-hour shift, these are his results:

NH: 4%

I: 10%

H: 86%

"It worked!" Fred shouts to a lone prairie dog poking his head out of a hole.

But did it? See, there may be a correlation between the awareness campaign and the increase in motorcyclists wearing helmets, but there's no proof of *causation*.

There could be lots of reasons that people are now wearing helmets. The price of helmets could have gone way down in the last six months. A new, less bulky style of helmet could have hit the market. It could be colder that day, and motorcyclists could have decided that a helmet would help keep their noggins warm. Or the whole thing could have been chance.

Point is there's more work to be done. Fred has helped collect the data, which is descriptive statistics. To find out why the data changed requires the use of inferential statistics. And in the end absolute causation may be difficult (or impossible) to prove.

Why does this even matter? If VROOMS assumes that awareness campaigns change minds, leadership might decide to spend 80% of its funds on these efforts. Is that a good use of money or not?

Ultimately, gathering data will probably create more questions than answers. And that's exactly what's supposed to happen. Researchers need a baseline and a way to monitor change along the way. Like other scientists, the good ones keep their assumptions in check and let the numbers do the talking—even if it says something completely different than they expected.

So say it with me: *correlation does not imply causation.* (Did you say that out loud? If not, go back and try again. It's important.)

This could very well be the biggest lesson for writers, editors, and publishers. Too often non-statisticians either forget or never learn this important fact, and that's led to a ton of bad reporting. A ton.

Even if an editor pressures you to suggest causation in your story, resist. Your reputation is on the line, along with the integrity of the research. None of us want to mislead our readers. Unfortunately this happens too often in the name of a snappy headline or great conclusion.

Do yourself a big favor: err on the side of caution here.

Reading a study or poll results

We've established this much: the sample—and how it was formed—is a big deal. It's also important to know what kind of statistical analysis was done. Was the point simply to gather data? Or did the researchers intentionally look for causation, using tools like *p-value* or *Bayes Theorem*?

That's where the actual study comes in handy. Remember, simply reporting study results from another news story or even a press release is not a great idea. Ask for a copy of the study itself.

Once you've done that, more than likely, you have a huge document on your computer or printer, full of charts, tables, and any number of confusing numbers and symbols.

Don't panic.

Remember, you don't need to be a statistician to understand the results of a study or poll. Here's how.

1. Do a little background research on the group or people who conducted the study or poll. If you suspect bias—for any reason—look for another source.
2. Read the summary first. Pay attention to your own questions. Look for information about sample size and methodology.
3. Look carefully at the summarized data, charts, and graphs. Be sure that the scales of the graphs are consistent and not misleading.
4. For a poll or survey, ask for the full list of questions (or at least the relevant section). If the pollster won't give you a list, you can't trust the poll. Period.
5. Find the margin of error or significance, which should be listed in the study itself. The margin of error is a percentage that expresses the amount of random sampling error in the results. If this value is large, say 15%, you shouldn't have much confidence in the results of the study. But if the margin of error is closer to 3%, it's probably safe to trust the results (all other things being equal).

6. Check the confidence interval or significance of that error. The confidence interval should be ±5%, while the significance is 95%. This means that the findings have a 95% chance of being true. (Subtract 5% from 100% to get 95%.)
7. Look for the sample size of the poll or survey. Consider how the sample was taken, and look for any possibility of bias.
8. Contact the researchers directly if you have any questions about how the data was collected or interpreted.

If everything looks kosher, you can feel reasonably certain that the study is reliable.

The significance of significance

Here's another little word of warning: make sure your story doesn't revolve around meaningless differences or changes. Take a look at this example:

In a poll, 45% of the respondents said they would vote for Mayor Jones. The results also showed that Commissioner Smith got 42%, and 13% of respondents were undecided. The margin of error is ±3 percentage points. That means Jones' actual percentages could be as low as 42% (45 – 3) or as high as 48% (45 + 3). Smith's actual percentages could range from 39% to 45%.

Now, it's really tempting to look at the ranges (created by the margin of error) and draw some conclusions. Smith could get 45% and Jones could get 42%, meaning Smith wins, or Smith and Jones could both get 45%. In other words, you might think, these two candidates are statistically tied. Not exactly.

In fact, if the study was conducted well, there is a greater than 5% chance that the candidate in the lead in this poll is not winning among the larger population. In other words, the numbers can't really say for sure which person is going to win that election. The poll results are too close.

In the end, the margin of error and significance say much more about how the poll was conducted than what we can expect to happen on Election Day.

It's best to leave significance to the statisticians. But there is one big thing to take away from this: avoid the term "statistical tie." It may seem like a great way to describe these situations, but in the end, it's best to simply say that the race is too close to call.

Polling questions

One thing that writers are good at is asking questions. And when you're reporting on a poll or survey, there's one thing you should make sure to do: ask for the questions.

There's a very good reason for this. Along with sampling, how the questions are worded has the biggest effect on the results.

So what happens if pollsters won't share all of the questions? Dump them. Transparency is critical. And when you do get the list of questions, pay close attention to the following.

Clarity. Read each question aloud. If you stumble over the wording, so did the respondents.

Double-barreled questions. Just like a reporter in an interview, a pollster sometimes asks two questions at once. But unlike reporters, pollsters only allow one answer: "Do you favor or oppose health care reform and strict price controls?" A yes answer may refer to health care or strict price controls or both. (And by the way, what does "strict" mean?)

Loaded questions. When there is some kind of assumption made in the question itself, it can be tough for the respondent to answer. Loaded questions are rarely found among reputable pollsters but are rampant in special interest groups. Some are obvious: "Do you favor or oppose the president's health care proposal, which will lead to lower costs?" (The assumption is that the president's health care proposal will lead to lower costs.) Others are less obvious: "How easy was it to find what you were looking for on our website?" (The assumption is that the visitor found what he or she was looking for.)

Leading questions. A leading question is one that suggests its own answer or reveals the bias of the pollster. "Do you believe murder is a sin?" might evoke a different response to a question on abortion than this lead-in: "Do you believe the government should stay out of people's personal lives?" Both of these are leading questions.

Can respondents answer the question? Whenever there's a big news event, pollsters ask what the effect will be. Examples include, "Will health care reform cost jobs?" and "Will the US attack Iraq?" Although these questions might help you figure out whether the public is aware of a policy issue, it tells you nothing else of meaning—the vast majority couldn't possibly know.

The center holds: mean, median, and mode

Readers don't want to wade through reams of raw data. Our job as writers and editors is to interpret the data for readers, so they can get a good sense of the story.

Besides, who has the column inches or bandwidth anyway?

This is where measures of center can be really useful. "Measures of center" give a general sense of where the "middle" of a bunch of data is. This helps people interpret and understand the data.

Three commonly used measures of center are mean, median, and mode. And if numbers tell a story, each of these measures tells a slightly different version, using the same data. So, it's really important to choose the measure of center that will relay the information most accurately and responsibly.

First up is mean. But let's get one thing out of the way before digging in. In terms of the numbers themselves (and the arithmetic used to find these values), mean is the same thing as average. Always and forever. No matter what anyone tells you, there is no difference.

That said, some statisticians use "average" to describe all three measures of center (mean, median, and mode), while others are content to use "average" and "mean" as synonyms. For our purposes, it's okay to consider mean and average as describing the same thing. But you might want to keep an eye out for slightly different uses.

The mean is the arithmetic average of a set of numbers. You're probably most familiar with this idea from school. To find your geometry grade, your teacher averaged the results of your tests, quizzes, and homework grades.

Tests: 97, 83
Quizzes: 73, 94
Homework: 100, 50, 80

If each of these is worth the same amount—or have the same *weight*—all your teacher did was add them up and divide by the total number of grades, like this:

$$97+83+73+94+100+50+80=577$$

$$577 \div 7 = 82.43 \approx 82$$

(Where did that 7 come from? Count the number of tests, quizzes, and homework; there are 7 grades in all.)

More than likely, your teacher didn't average grades this way. Why? Because tests probably counted more than quizzes, which counted more than homework. This is called a *weighted average*.

Let's say that the tests are counted as 50% of the final grade, quizzes are 30%, and homework is 20%. There are lots of ways to figure this out, but here's one.

First find the average of each of the different kinds of grades:

Tests: $(97 + 83) \div 2 = 180 \div 2 = 90$
Quizzes: $(73 + 94) \div 2 = 167 \div 2 = 83.5$
Homework: $(100 + 50 + 80) \div 3 = 230 \div 3 = 76.67$

Now, multiply each of these averages by their weights and add:

$$(90 \cdot 0.5) + (83.5 \cdot 0.3) + (76.67 \cdot 0.2)$$

$$45 + 25.05 + 15.334$$

$$85.384$$

That means your grade for that quarter (rounded to the nearest ones place) is an 85.

The mean or average is very common in calculating grades or physical attributes, like weight or height. In reporting, it should be used with caution. That's because in the data set, a really large or really small number—or an *outlier*—can throw off the mean. A really high value will pull up the mean, while a really low number will pull it down. In those situations, the average may not be a good representation of the data.

This is where the median can be very useful. Often called *typical values*, the median is used with dollar values, like income or house values.

The median is the middle value of a set of numbers. Half of the values are above the median, while half of the values are below it. If you have an odd number of values, the median will be one of the values in the data set. But if you have an even number of values, the median is the mean of the two middle values.

So what's the first step in finding the median? Getting all of the data in order from smallest to largest (or largest to smallest).

Senator Jennifer Finker-Lawson has asked her staff to pull together some numbers for an upcoming speech at the Women in Economy Conference. Having put a great deal of effort into bolstering the local economy, Sen. Finker-Lawson is sure that the median salary for workers in her district is higher than last year. She'd like to tout these numbers from the podium.

Her staff collects the average weekly wage from 7 important industries:

government (federal, state, and local):	$1,207
education:	$933
health care:	$1,045
professional services:	$1,370
manufacturing:	$1,324
information and technology:	$1,509
leisure and hospitality:	$385

To find the median, they put the results in ascending order. (Descending order would have worked, too.)

385, 933, 1045, 1207, 1324, 1370, 1509

Right away, the staff notices an outlier. The leisure and hospitality industry earns far less per week (on average) than the other industries. Finding the median is a good idea, because the $385 would simply pull down the mean, misrepresenting the data.

The number exactly in the center is $1,207. (There are 3 numbers less than this one and 3 greater than this one.) So the median average weekly salary for these 7 industries in Senator Finker-Lawson's district is $1,207. Last year, this value was $1,189, so there is an increase.

But just because the number rose at the same time that the good Senator was putting some effort into increasing these values doesn't mean that Jennifer caused the increase. There might be a correlation, but that doesn't mean there's causation.

Of course that doesn't stop any politician from taking credit—especially in an election year. Still, reporters should be careful not to perpetuate that fallacy.

A word of warning: if you're writing for the general public, take a moment to explain what a median is. While folks in economics, real estate, and other money-related fields will likely have no trouble with the idea of median, the average grownup probably needs a little refresher.

The last measure of center is the mode, or most frequently occurring value. To be honest, you're not going to use this measure very often, but there are times when it's useful. Here's an example.

Gwen is writing a story about Acme Widget Company. The owner of the company reported that the median salary is $400 per day. There are 10 line workers, 7 assistants, 3 managers, and 1 owner. Gwen is a bit suspicious, so she asks some follow-up questions.

Good reporter, that Gwen. Turns out that the line workers make $100 per day, the assistants make $400 per day, the managers make $900 per day, and the owner makes $1,900 per day. That's a pretty big difference!

Gwen first checks to see if the salaries really do have a median of $400. She arranges all of the salaries in order from least to greatest and then chooses the middle value:

100, 100, 100, 100, 100, 100, 100, 100, 100, 100,

400, 400, 400, 400, 400, 400, 400, 900, 900, 900, 1900

There are 21 employees (salaries), so the median is the 11th number in the list or $400. So the owner of the company reported the correct median.

Gwen wonders about the mean next. She adds all of the salaries and divides by 21 or the total number of salaries.

$$100 + 100 + 100 + 100 + 100 + 100 + 100 + 100 + 100 +$$
$$100 + 400 + 400 + 400 + 400 + 400 + 400 + 400 + 900 +$$
$$900 + 900 + 1{,}900 = 8{,}400$$

$$8{,}400 \div 21 = \$400$$

So the mean and median are the same thing. That's interesting, but what really has Gwen thinking is this: there are so many more line workers than managers and the owner is only one person. The differences in the salaries are really striking. Thinking about the number of people in each kind of position has Gwen wondering: what's the mode?

$100	10 employees
$400	7 employees
$900	3 employees
$1,900	1 employee

The $100-per-day salary occurs more often than any other salary on the list, so the mode is $100. It might be more revealing to share not only the median—because the values are in dollars—but the mode, as well.

A little warning: don't confuse a mode with *most*. The mode is the most frequent category, not necessarily a majority of the whole. For example, in this list:10, 10, 10, 9, 8, 7, 3, 3, 1, the mode is 10 because it is the most frequent category. But it isn't most of the numbers—it's only 3 of 9.

Home, home on the range

Range is another value that can be very useful when you have a big list of numerical data. This is *not* a measure of center. Instead range is simply the difference between the smallest and the largest values in the data set.

In Gwen's story about Acme Widget Company, the range of salaries really is interesting. Take a look:

The largest salary is $1,900
The smallest salary is $100
The range is $1,900 – $100 = $1,800

That's a pretty significant range!

Pretty pictures

Of course there are lots of ways that data can be displayed as charts, graphs, and even images. And with the publishing industry running on a shoestring these days, editors are counting on writers to at least suggest these options—if not provide rough sketches of them. One way to outshine other writers is to offer these extras.

Then there are the bloggers and self-publishers among us. Without an editor or design team behind us, we're left to create pie graphs and tables all by our lonesomes.

Graphs should never replace reporting on important information about polls and surveys. Somewhere in the story or in a caption of the graph, you should be sure to mention where the information came from, what the sample size was, and how the data was collected. Assure your readers that you are reporting on responsible data.

Here's a quick rundown of the types of statistical graphs and when to use them.

Pie Graph

Also known as a circle graph, the pie graph shows parts of a whole. Each section of the graph—or slice of pie—represents a different category in the data. The graph below shows the results of survey about pie. (Get it?)

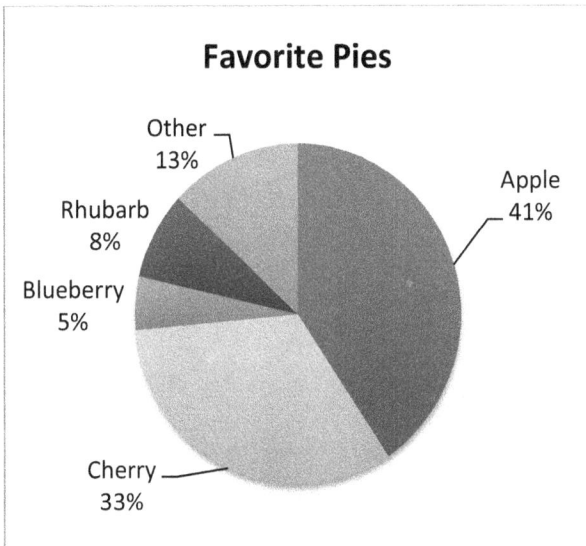

Favorite Pies

Other 13%
Rhubarb 8%
Blueberry 5%
Cherry 33%
Apple 41%

The key thing to notice is that if you add all of the percentages, you get 100%.

41% + 33% + 5% + 8% + 13% = 100%

If the percentages do not add up to 100%, there could be two things going on. First, the percentages may have been rounded, making the total 99% or 101%. That's generally a-okay. (Just be sure to mention that you've rounded the individual percentages. One idea is to add a note at the bottom: "Percentages total more than 100% due to rounding.") But if the sum is, say, 60% or 130%, there's a big problem with the data.

In the pie graph above, there's a lot of information that you *can't* see. For example, there's no way of knowing how many people were sampled, since you have only the percentages. This data can be shown in the graph, or it can be left off. And that's generally true for all types of statistical graphs.

Bar Graph

With a bar for each category of data, a bar graph is used to compare the results among different groups. This works very well with data that is divided into subgroups.

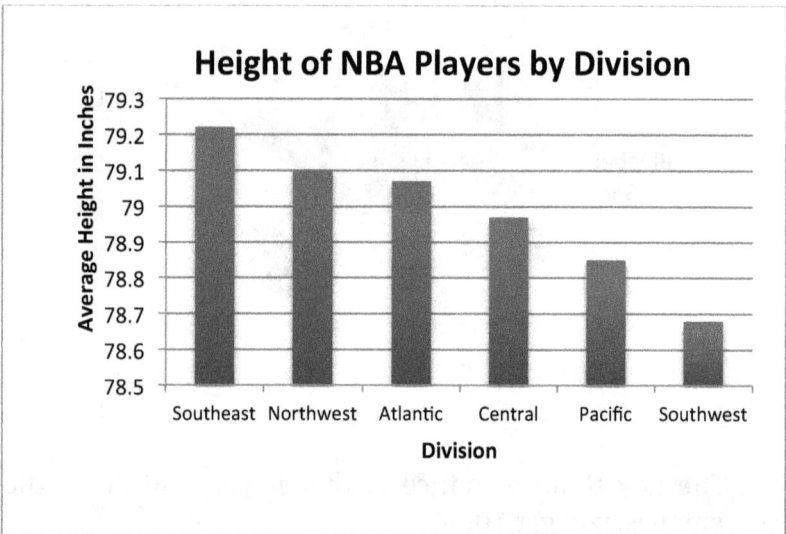

The scale of the bar graph is critical. At first glance, this graph on the previous page seems to show a big difference in the heights of players in the Southeast Division and players in the Southwest Division. But if you look closely at the "Average Height in Inches," you can see that the scale is really tiny—only 0.1 inch. Rescaling this graph gives a more accurate description of the data.

Height of NBA Players by Division

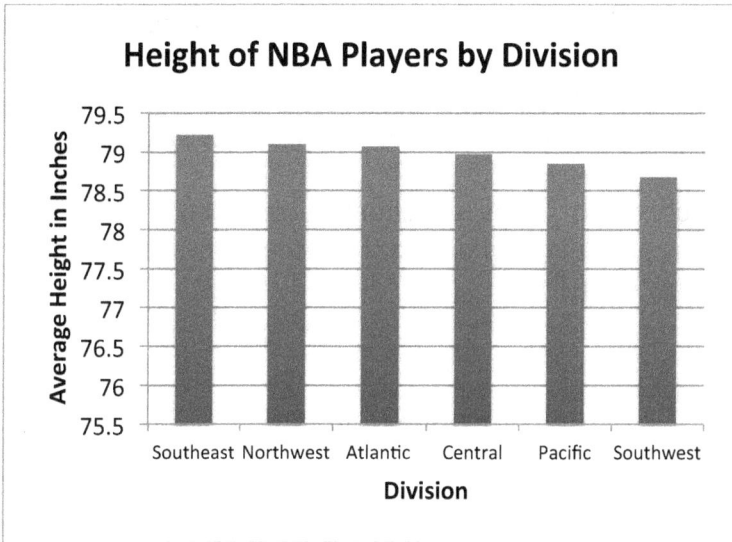

Line Graph

A line graph is used to show continuity within data. Data points are graphed, and a jagged line connects those dots.

Temperatures on October 11, 2013

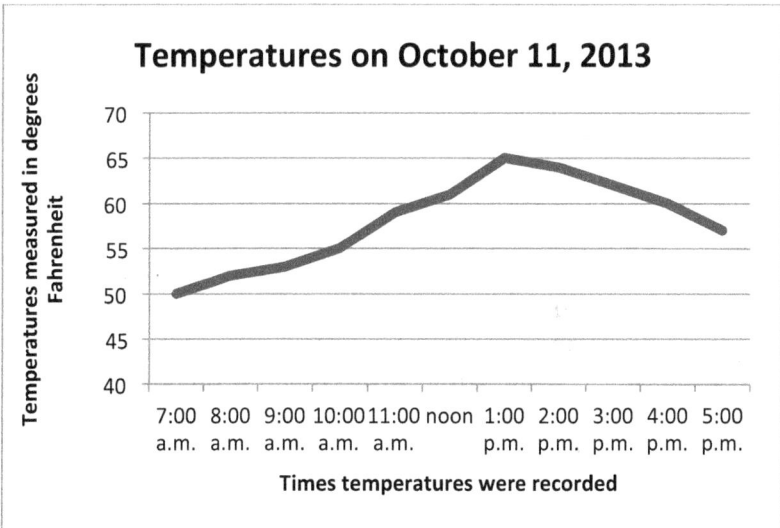

Continuous data can take any value within a range. For example, if you measure the air temperature in small enough increments of time, you can be sure that between 2 measurements, the temperatures are dropping or rising or staying the same. The changes in temperature follow a predictable pattern.

But a line graph is frequently misused. For example, this line graph is intended to show the monthly profits for a company over 1 year.

Monthly Profits

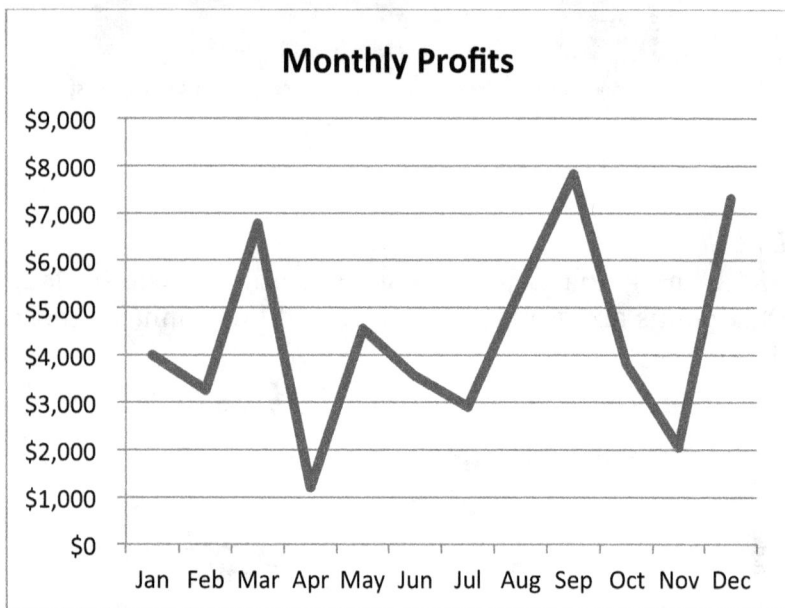

This data is not continuous. In the line graph on the previous page, the data only shows what the profits are at the end of each month. There is only 1 measure per month, not a continuous measurement from the beginning of the month to the end. In fact, a bar graph is more representative here.

Histogram

It's pretty likely that you have a) never heard of a histogram, or b) conveniently forgotten what it is. They look a lot like bar graphs, but they are very different.

Like line graphs, histograms display continuous data. For that reason, there is no space between the bars. The ranges of values are called *classes* and listed at the bottom. What is being measured is *frequency*—or the number of members in each class. The taller the bar, the more members in the class.

This histogram shows the distribution of IQ scores among a sample of sixth graders.

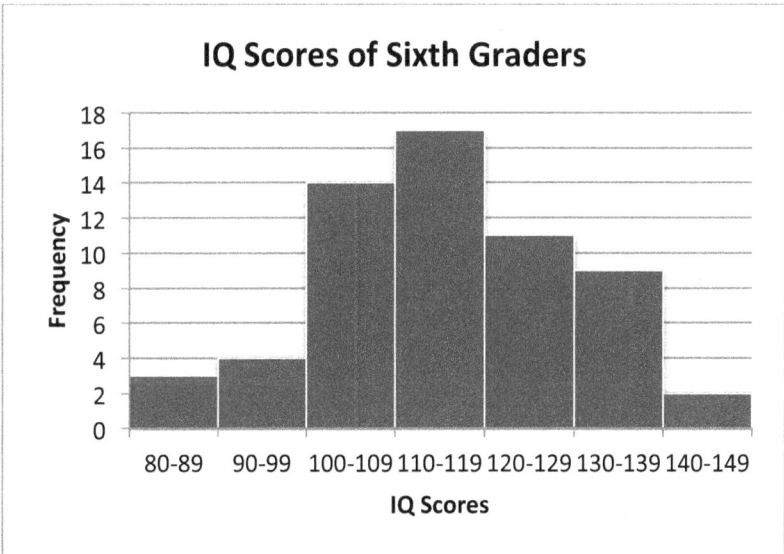

IQ Scores of Sixth Graders

(Histogram: x-axis labeled "IQ Scores" with classes 80-89, 90-99, 100-109, 110-119, 120-129, 130-139, 140-149; y-axis labeled "Frequency" from 0 to 18. Bar heights: 80-89 = 3, 90-99 = 4, 100-109 = 14, 110-119 = 17, 120-129 = 11, 130-139 = 9, 140-149 = 2.)

This data in this histogram should make logical sense. Most of the IQ scores are clustered around the middle ranges, 100–109, 110–119, 120–129, and 130–139. Many fewer students have low or really high scores.

Notice anything else? If you were to draw a curve that outlined the shape of this histogram, it would look like a bell. That's the infamous *bell curve* that scientists love to talk about. Simply put, if the data is distributed as a bell curve, most of the results are in the middle ranges, while fewer study members are on the edges. Another word for this is the *normal curve*, and it's a pretty big deal in statistical analysis. That's because a lot of data follows the normal curve. In fact, many calculations about probability can be approximated using the normal curve. This is what allows researchers to draw inferences about the data they've collected.

Scatterplot

Lastly, there's the scatterplot. In this graph, the data is plotted using points. It's used to demonstrate correlation between 2 events or characteristics. This scatterplot displays the relationship between median home prices and median salaries.

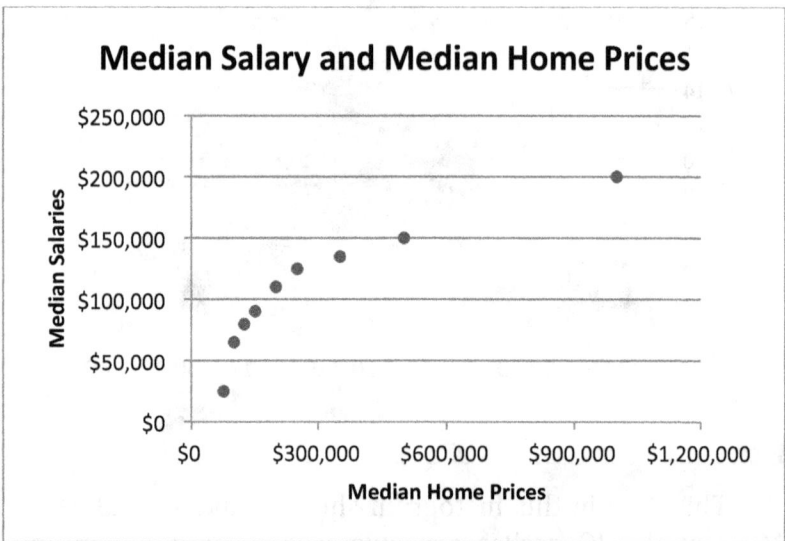

Median Salary and Median Home Prices

The scatterplot can also be used to predict values that are not graphed. For example, based on this data, what is the median salary when the home price is $800,000?

The data does create a pattern, but it's not enough to just connect the dots into a curve or jagged line. That's because you can't really predict what the curve or jagged line will look like at certain points. But there is something called the *line of best fit*.

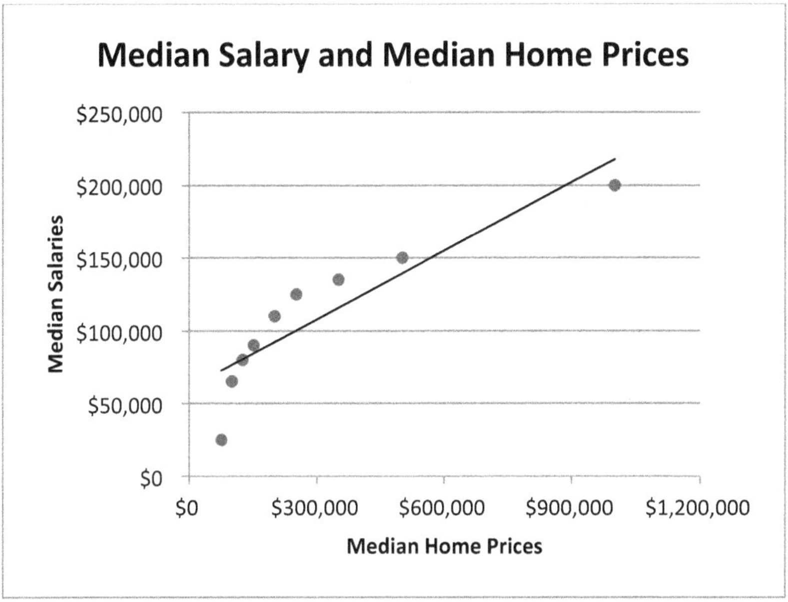

The process of finding the line of best fit is called *linear regression*. In other words, you want to draw a straight line that either intersects or gets very close to as many points on the graph as possible.

The point is to be able to predict other points in the data set. For example, in the graph above, there is no value at the $800,000 median home price. But looking at the line of best fit (drawn in the second graph), you can predict that for neighborhoods with that median home price, the median salaries are somewhere around $175,000.

Keep in mind that correlation does not imply causation. Just because there's a demonstrated correlation between median salaries and median home prices does not mean that one causes the other. (Getting tired of hearing that, yet?)

Want to see more examples of graphs—in particular, problematic pictures? Visit www.mathforgrownups.com and click on "Writers" in the navigation bar. You can even learn how to make these graphs all by yourself!

Warning: do not pretend

Trust me on this. I've sat through many a stats class—and even taught and written a few high school courses. Stats can screw up your brain.

Not everyone feels like they've fallen down Alice's rabbit hole when they're asked to evaluate a sample or determine if results are reliable, but many of us do. Even the most agile mathematician will tell you—statistics can be a very different beast.

Here's some advice:

Check Yourself

Don't just check to see if the average you got from your source is correct. Check your own assumptions about that average. Did you expect it to be higher? Lower? Is an average the best way to describe the data?

For example, a source tells you that the average salary at his company is $94,340 per year. Your gut tells you something's a little off here. Seems you remember reading that the owner was making $1.5 million per year. So you ask him for a list of all salaries by position, as well as the number of employees at each position.

Turns out you were right to dig a little deeper. The owner's salary is what we call an outlier—an extremely large (or extremely small) value compared to the rest of the data. In that case, an average isn't such a great statistic. Instead, it's better to use the median.

The upshot is this: just because you got a number doesn't mean you've got the story. Treat numbers with great care, to resist making mistakes in how you—and your readers—interpret them.

Skip the Fancy Stuff

Again, unless you've spent some time learning and using basic statistics processes, leave the *p-values* and *standard deviation* to the experts. There's no need to get into the nitty-gritty.

But if you do want to dig deeper, get some professional help. Take a class, dig into some online videos, or read a stats book. Just don't start using a *p-value* until you're certain you know what you're doing with it.

Make Friends with a Statistician

If you write in a field that requires a great deal of stats—like health or business—ring up the math department at your local university or community college. These teachers really do understand how confusing stats can be. They're extremely frustrated when they see poor stats in news articles. And that means someone is eager to help you. With the right person, you can ask specific questions about the math or even send over sentences, grafs, and entire articles for a quick look-see.

I guarantee there is some lonely stats teacher out there who is really happy to help you understand his or her field better.

Here's the bottom line. When it comes to stats, don't get cocky. Check, double check, and triple check. Your sources, editors, readers, and conscience will thank you.

Part 2 — Get Published

WOO-HOO! YOU'VE HIT ON the perfect book or story idea. It's fresh, it's fascinating, it's phenomenal.

The only thing left to do is sell the darned thing.

Perhaps the most jarring lesson for aspiring writers is the one where we figure out that writing the story or book is *not* the first step. Nope, we've got to actually pitch our ideas to overworked editors who don't have a clue who we are and whose inboxes are overflowing with ideas from other writers.

If you're really new and haven't gotten this far in the writing-for-a-living journey, let me be really clear: most of the time, stories and books are assigned based on a query or proposal. And that's a good thing. Who wants to report and write a 1200-word story from start to finish and *then* shop it around from *Good Housekeeping* to *Glamour*, only to find out that no one is interested?

A query or book proposal should sell your idea. And that means you need some good, solid information—including statistics—that demonstrates why this is a good idea, why it should be published now (and in a particular publication), and why you are the best person to write it.

Having a few math tricks up your sleeve can up your query-to-publication ratio.

(See what I did there?)

Editors and agents get hundreds of pitches a week, and with really great assistants, they see only the ones that pass muster. To get their attention, writers have to answer three very important questions:

Why us?

Why now?

Why this writer?

All of these questions can be answered using a little bit of math. Sure, the numbers aren't going to tell the whole story, but giving some concrete data can help convince the editor that you and your idea are worth investing in.

That's why impressive queries, pitches, and book proposals have some good, solid data behind them. Numbers explain why the story is a good match for a particular publication. Numbers describe why the story is important now. And numbers illustrate why *you* are the best person to get this story out there.

If you're not including some data in your query or book proposal, you're missing big opportunities to sell your idea.

These questions aren't just for writers in the traditional publishing industry. For the independent folks among us, it's a good idea to make a plan before launching into the self-publishing industry. Just because it's easy to get an e-book out to the masses doesn't mean you should.

These days, aspiring authors also have to think about something called *platform*. With the enormous changes in the publishing industry, agents and acquisitions editors are particularly picky these days. Marketing and press budgets have been slashed. Authors are expected to take on much bigger roles to be sure that their books are successful. And self-published authors are completely responsible, soup to nuts.

Your platform is one way a literary agency or publishing house determines whether you're worth the risk. Here's the really cool thing: even if your platform looks really small, some crafty statistical analysis can reveal that it's very deep. That's the stuff that separates run-of-the-mill authors from the real success stories.

A little bit of math can actually help you figure out how to sell a story, build your platform, or even decide whether independent publishing is worth it for you.

Note: For the most part, I'll focus on book publishing, but much of what is said here can be applied to magazine or newspaper queries. It's not necessary for a writer to have a platform in order to write articles, but it can be helpful. One way to think of the story query is like a mini– book proposal.

What are you waiting for? Let's get this party started.

Chapter 5 — To Market, to Market!

ANY GOOD WRITER WILL tell you that at least 50% of the job is sales. Surprised? Well, unless you *want* to be a starving artist or you're okay with simply journaling or blogging your greatest ideas, research, and reporting, getting your stuff out there to the masses requires some help from others.

And those others have big gatekeepers. With a limited news hole and a shrinking public attention span, editors, agents, and publishers have to be pretty picky about whom they'll take a chance on. They're taking calculated risks. And the first one is this: will this idea actually sell?

With the plethora of self-published e-books clamoring for attention—or being overlooked by the masses—even independent publishing has some pretty tall obstacles. Having a good idea of who will read your book and how they'll find out about it is a great first step in any publishing process.

That's where market research comes in. In fact, in a traditional proposal, two sections include significant market research: your promotion plan (which includes an assessment of who will buy your book) and the competitive analysis (which looks at other titles in your subject area).

There are two parts of doing market research: gathering the information and then crunching the numbers to show why your book should be published.

Debbie has a great idea. She'd love to publish a book detailing the 12 most important works of fiction from the last century. Each chapter will include a brief outline, an analysis of the novel, and a list of discussion questions.

"It'll be like a written book group!" she cries.

She's already got the fiction picked out, and with her dissertation in 20th century American literature—comparing and contrasting the works of Eudora Welty and Zora Neale Hurston—a lot of the research is out of the way.

Debbie has already decided she's not going the independent publishing route. She has big plans for this book—and the ones that follow it. She wants to find an agent and a big publishing house.

Girl's got dreams ain't nobody going to snuff out.

She borrows *10 Steps to a Perfect Book Proposal* from the public library and studies it cover to cover. Then she blocks out a long, snowy weekend to sit down and perfect her pitch.

The outline and overview are easy. Those details come from her head, and she's sure they'll wow anyone in the publishing industry. The next step is to dig into some market research. She hits the Internet.

Debbie starts out looking at secondary research or data that someone else has collected. She's decided that her audience includes well-educated and relatively well-to-do women. These women have college degrees, but they're older—maybe in their 40s or 50s—and they want to learn a bit more about modern, classic literature. At this point in their lives, they're interested in merging their leisure time with a little bit of self-improvement.

Most importantly, they belong to book groups.

With this information, Debbie starts Googling. First, she heads to the US Census website. She figures that this is the best place to find out the number of women in the country between the ages of 40 and 50 years.

She downloads the latest spreadsheet detailing Age and Sex Compilation and highlights the cells with the data she needs.

Table 1. Population by Age and Sex: 2011
(Numbers in thousands. Civilian noninstitutionalized population[1])

Age	Both sexes		Male		Female	
	Number	Percent	Number	Percent	Number	Percent
All ages	306,110	100.0	150,643	100.0	155,466	100.0
Under 5 years	21,265	6.9	10,869	7.2	10,396	6.7
5 to 9 years	20,870	6.8	10,667	7.1	10,203	6.6
10 to 14 years	20,020	6.5	10,237	6.8	9,783	6.3
15 to 19 years	20,886	6.8	10,691	7.1	10,196	6.6
20 to 24 years	21,525	7.0	10,960	7.3	10,565	6.8
25 to 29 years	21,382	7.0	10,917	7.2	10,464	6.7
30 to 34 years	20,202	6.6	10,067	6.7	10,135	6.5
35 to 39 years	19,255	6.3	9,542	6.3	9,713	6.2
40 to 44 years	20,587	6.7	10,172	6.8	10,415	6.7
45 to 49 years	21,989	7.2	10,800	7.2	11,189	7.2
50 to 54 years	21,965	7.2	10,695	7.1	11,270	7.2
55 to 59 years	19,554	6.4	9,499	6.3	10,055	6.5
60 to 64 years	17,430	5.7	8,447	5.6	8,983	5.8
65 to 69 years	12,160	4.0	5,600	3.7	6,561	4.2
70 to 74 years	9,254	3.0	4,242	2.8	5,012	3.2
75 to 79 years	7,088	2.3	3,065	2.0	4,023	2.6
80 to 84 years	5,719	1.9	2,370	1.6	3,349	2.2
85 years and over	4,957	1.6	1,804	1.2	3,153	2.0
Under 15 years	62,155	20.3	31,773	21.1	30,382	19.5
15 to 17 years	12,761	4.2	6,570	4.4	6,191	4.0
18 to 20 years	12,515	4.1	6,377	4.2	6,138	3.9
21 to 44 years	98,562	32.2	49,402	32.8	49,159	31.6
45 to 64 years	80,939	26.4	39,441	26.2	41,498	26.7
65 years and over	39,179	12.8	17,081	11.3	22,098	14.2
Median age	36.8	(X)	35.5	(X)	38.1	(X)

(X) Not applicable
[1] Plus armed forces living off post or with their families on post.

SOURCE: U.S. Census Bureau, Current Population Survey, Annual Social and Economic Supplement, 2011.

Then she adds the numbers in those cells:

$$10,415 + 11,189 + 11,270 + 10,055 = 42,929$$

"Huh," Debbie says, chewing on the end of her pencil. "That's not very many people."

Glancing to the top of the chart, she notices what she missed. In her excitement, she completely forgot to consider how the data was included in the table. But right there, in black and white, it says: "Numbers in thousands."

So the total is actually 42,929 *thousand*. Debbie knows this means she only needs to add three zeros to the end of the total: 42,929,000 or nearly 43 million women between the ages of 40 and 60. That makes much more sense.

Just for the heck of it, Debbie decides to find the percentage of the population that fits this demographic. To do this, she needs the total number of women in their 40s and 50s (42,929,000), and she needs the total number of people in the US (306,110,000).

When Debbie set out to find the total number of people in the US, she could have gotten this information from a variety of sources. But she chose the results of the Census, because that's where she got the number of 40- to 60-year-old women. It's important to compare apples to apples. Even if there is more current information out there, stick with the data that comes from the same source.

Now she divides:

$$\frac{42,929,000}{306,110,000} = 0.14 = 14\%$$

Now what about education? It wasn't realistic to think that every single one of these women would be interested in reading a book about great literature. But she figured that women with college degrees might be.

She heads back to the Census website to see what it had to offer. Pretty quickly, she finds an "Educational Attainment" table that has details by sex. Deciding that she wants to include those with a Bachelor's degree or higher, she adds up the data from those cells:

$$21,146 + 9,062 + 1,155 + 1,162 = 32,525$$

Again, the data is in thousands, so Debbie needs to add some zeros. She finds out that according to the US Census, there are 32,525,000 women in the country with college degrees (not counting associate degrees). And just in case she decides she needs it, Debbie finds the percentage:

$$\frac{32,525,000}{306,110,000} = 0.11 = 11\%$$

A note indicates that the data only includes women who are 25 years and older. She thinks for a moment about how she might be able to parse this data further: can she find out how many women between the ages of 40 and 60 have college degrees? Well, maybe, but it's a little tougher. She decides that having just these two numbers is good enough.

Within 30 minutes, Debbie has found and tabulated two numbers for her market research. Does this mean that Debbie can expect to sell 43 million copies of her book? Nope? What about 32.5 million? Probably not.

In fact, really general numbers like these are not necessarily all that useful in a book proposal. If Debbie stops here, she'll look like a real novice.

Good thing there's another bit of information she wants to track down: book groups. The US Census isn't going to help her out this time. Instead, she needs to find a reputable organization that has done this kind of research. She Googles "How many book clubs are there in the US" and finds several articles on the subject.

One says that 5 million Americans belong to book clubs. Another puts the estimate at between 20 and 25 million. That's a huge difference and there are no citations for these, so Debbie keeps looking. She really needs a primary source that she can count on.

She hits a major brick wall. She can't find any organization that does a good job of counting the number of book clubs in the country. What she does find are some larger online book clubs. So she starts taking notes.

First, there's Oprah. Her website reports 2 million book club members, many of whom are reading the classics.

Next up are the online book review aggregators. Once site lists 682 online book groups that are devoted to classic literature.

But wait! There's more! The blogging platform Tumblr recently unveiled a book club, and in with its first book, logged 3,175 notes or mentions. The National Endowment for the Arts offers community organizations grants for launching community-wide book clubs. This program, called The Big Read, offers support materials for 30 different books, most of them classics that Debbie would like to include in her book. Next year, 77 organizations will launch programs with The Big Read.

After 30 minutes furiously searching the Internet, Debbie's head is starting to swim. Clearly this research is going to take a long time, but she feels confident that she can track down some great figures that support her book idea.

But this market research demonstrates that there's a big, big pool of people who are in the book's target market. Even more importantly, there's a captive audience that's relatively easy to reach—book group members. Those are pieces of the puzzle that an editor or agent needs to have.

But if Debbie can't clearly identify these numbers, she might need to adjust her expectations a bit. Perhaps her book is only going to work for a niche audience. In that case, a smaller publisher could be her best bet. And that's exactly what she decides to do.

Even with that decision made, Debbie can't stop at audience, though. She remembers the old adage that there are no new ideas. There are other books like the one she wants to write already in the marketplace. Some are competitive titles—books that are similar in subject and approach. Others are complementary titles—books that further her idea in a different way.

Debbie needs to know how well these books are doing.

Why? Because the publisher will want to know. Book publishing is a lot like movie making. When something does really well, publishers want to replicate that success. So we see a plethora of zombie movies and books that detail a yearlong experience (*The Happiness Project, Julie & Julia,* and *The Year of Living Biblically*). In this case, it can be advantageous to be part of the crowd.

But how do you find out how well books like yours are performing? The Internet is useful here.

Debbie types in www.amazon.com and starts searching for books about book groups. The results fall into three categories: "no relevance," "competitor," and "complementary." For each title, Debbie jots down important information—like publication date, publisher, edition, printing, and price. Then she starts looking at the Amazon ranking numbers.

The sales rank is based on a carefully guarded and complicated formula. Still, plenty of folks have gathered information suggesting that the rank depends on several variables: number of sales (or giveaways) in a given time period, consistency of sales, and price (yes, price). No matter what, this number is the best measure that most writers have for assessing how well a book is doing.

In Debbie's research, she finds that most books about book groups, including novels and how-tos, are hovering at a sales rank of around 50,000. (Not an impressive number. Debbie is feeling much better about her plan to go with a smaller publisher.)

She can list each sales rank or find a median or mean. (Debbie remembers that when there's an outlier—a really big or really small number in the list—the responsible thing to do is find the median.)

Meanwhile, there are complementary books to consider. For example, if she's got a chapter about *Their Eyes Were Watching God*, it could be interesting to find the Amazon sales rank for that book as well. Bingo! Number 52 in African American Literature and #60 in Classic Literature. Those little digits are going to be *very* helpful.

She can argue that since Zora Neale Hurston's book is doing very well in its subcategories, there is interest in the literature. Furthermore, sales can be boosted if her book is paired on Amazon with the classics that she covers.

But in the end, Debbie's disappointed with what she's discovered. She was hoping to find competitive titles that are selling well, as well as a group of easy-to-reach readers who are just waiting for her book to come along.

After wallowing a bit, Debbie gets serious again. Even if she didn't find what she wanted, she can still write and publish her book. Her plan is simple: she'll shoot out proposals to smaller publishing houses and consider self-publishing. Meanwhile, she'll do some additional research on the market she wants to reach. Perhaps if she builds her platform, she could entice a larger publishing house or at least sell more books—regardless of how they're published!

Words to the wise

As you've heard again and again, not all data is created equal. Just because it's on the Internet doesn't make it true. And slapping some numbers in the book proposal is not enough.

Go to reputable sources for data. That includes the US Census, trustworthy research organizations, nonpartisan government agencies, and well-considered nonprofit agencies.

Really general information—like the number of people in your target audience—can be interesting but is not enough to sell the book. Drill down to much more specific audiences as much as you can. Get creative and use your research skills to find details that will really wow an editor, agent, or publisher.

Chapter 6 — Building a Platform

ASK A WRITER ABOUT his platform, and she'll likely roll his eyes into the back of her head. But for an agent or publisher, platform is the key to a successful book.

It's not only a big part of the book deal—but a huge consideration for independent authors (they need to sell their books, too, and they don't have a publisher to help out). When an agent or publisher asks about platform, she's really asking, "How many potential buyers can this author reach easily and quickly—through social media and other outlets?" Like most aspects of publishing, platform can be measured.

It's easy to get caught up in the game of "how many." Digging a little bit deeper, however, will give a truer picture of your platform. In other words, if your blog posts reach only 1,000 people each month (instead of the 15,000 that an agent might like to see), don't panic. There may be a way to demonstrate that your platform is actually much larger than the cold, hard data seems to show.

Emma has been covering the Supreme Court for more than 20 years. Her articles have earned her great adulation from fellow journalists and readers.

But she's a bit tired of the Supremes. Sure, the big, juicy cases are interesting, but Scalia's acerbic sense of humor is wearing on her, and the liberal worship of Ginsberg is way too much to stomach on most days.

What's got Emma's interest right now is a legal thriller that's boiling around in her brain. It might be time to try some fiction.

She thought of writing under a pen name, but with the publishing industry the way it is now, she knows that a solid platform is going to be important in landing the deal and making the sales. She doesn't need to make a million dollars, but if she's going to write the darned thing, she sure as heck wants lots of people to read it!

So after writing and editing a draft, Emma puts on her reporter cap and starts crunching some numbers.

First, she considers her blog, "Stop! In the Name of Law." She started blogging about a year ago, posting a couple of times each month. She's never looked at the stats, so she clicks into Google Analytics and takes a look.

There are *a lot* of numbers and charts on that page! After digging around, Emma notices a nice summary of monthly visitors, unique visitors, pageviews, pages per visit, average visit duration and something called "bounce rate."

Visitors	1,349
Unique Visitors	907
Pageviews	3,913
Pages per visit	1.67
Average visit duration	2:45
Bounce rate	62.9%

Only 1,349 visitors each month? And some of them are repeat customers? Emma's been in the media biz long enough to know that (a) these numbers aren't going to get her a nice book deal, and (b) if people don't know that a blog exists, they'll have a hard time finding it.

She digs around a little more and finds that most people get to her blog through searches. They search her name, the names of Supreme Court justices, key words from certain cases, and some really odd things, like "sexy judge." Ignoring the last category, Emma comes to a quick, rough conclusion: even without any promotion, her blog is reaching some folks. If she could ratchet up her posts and let people know about them, she might see a big increase in these numbers.

Emma has a strong Twitter and Facebook following. Her Twitter handle @supremereporter has 23,477 followers and her Facebook page (not her personal account) has 7,838 "likes." But like her blog stats, there are some charts and totals that she wants to look at more closely.

See, social media is more like a spider web than a ladder, Emma remembers. (She stole that metaphor from a colleague, who did a piece about viral YouTube videos last year.) When people retweet or comment or repost, the message is getting out to a much larger group than her own followers.

So Emma starts digging around further. She uncovers the following information from the last month:

Twitter followers:	23,477
Tweets:	42
Retweets:	12
Favorited tweets:	5
Facebook page likes:	7,838
Facebook posts:	8
Facebook post likes:	43
Facebook comments:	23

Even with her notoriety among Supremes junkies, Emma knows that she has some work to do before she can convince a publishing house to take a risk on her. Luckily, that work is all very doable—she resolves to ratchet up her social media interaction so that she can get some numbers that matter.

After a month, Johnson has some new numbers. She puts them in two tables, so she can keep track.

Blog:

Measurement	Month 1	Month 2
Visitors	1,349	2,471
Unique Visitors	907	1670
Pageviews	3,913	6,245
Pages per visit	1.67	2.02
Average visit duration	2:45	2:53
Bounce rate	62.9%	59.4%

Social Media:

Measurement	Month 1	Month 2
Twitter followers	23,477	30,018
Tweets	42	71
Retweets	12	29
Favorited tweets	5	11
Facebook page likes	7,838	8,410
Facebook posts	8	84
Facebook likes	43	289
Facebook comments	23	104

The experts were right. With a little bit of effort—mostly on the part of a virtual assistant she hired—those numbers went up. But not by a whole lot. Just about 2,500 visitors to her blog each month isn't going to wow anyone.

But Emma has a cool trick up her sleeve. She notices that her unique visitors and retweets nearly doubled. If she can demonstrate a sharp increase over just one month, the numbers might look much better. So she decides to find the percentage change of many of the categories.

Emma remembers that percentage change requires two values: the original number and the new number. To find the percentage change, she needs to subtract the old value from the new one and divide by the old one.

(new value – old value) ÷ old value

She decides to look at blog visitors first:

$$(2,471-1,349)\div1,349$$

$$1,122\div1,349$$

$$83\%$$

So in a month, the number of visitors to her blog increased by 83%. Not bad. Emma decides to calculate the percentage change for several of the other measurements. And her table ends up looking a lot like this:

Measurement	Month 1	Month 2	Change
Visitors	1,349	2,471	83%
Unique visitors	907	1,670	84%
Pageviews	3,913	6,245	60%
Pages per visit	1.67	2.02	21%
Average visit duration	2:45	2:53	
Bounce rate	62.9%	59.4%	

The last two measures, average visit duration and bounce rate, don't look like the other numbers. Because of that, calculating a percentage change doesn't seem like a great idea, so she leaves those blank and moves to the next set of data—social media stats:

Measurement	Month 1	Month 2	Change
Twitter followers	23,477	30,018	28%
Tweets	42	71	69%
Retweets	12	29	142%
Favorited tweets	5	11	120%
Facebook page likes	7,838	8,410	7%
Facebook posts	8	84	950%
Facebook likes	43	289	572%
Facebook comments	23	104	352%

All of these are positive changes, which makes perfect sense: from last month to this, all of her metrics went up. Looking further, Emma notices that some of the measures are a little harder to move than others. Those are the ones with the lower percentage change, like "pages per visit" on her blog, Twitter followers, and Facebook page likes.

She also notices a correlation between increasing her Facebook posting frequency and getting more likes and comments. When she posts more, she gets more "likes" and comments. Same for Twitter: when she tweets more often, she gets more retweets and favorites.

Emma knows that this is not necessarily a cause-and-effect situation. But there's clearly a relationship between being more active on social media and getting noticed. She suspects that if she tweeted random letters or numbers, her retweets might plummet—or they might skyrocket, as her followers wonder what the heck is wrong.

But how should she measure the change in "average visit duration" and "bounce rate" on her blog? She could find the percentage change for the first value, but that seems a little tricky. Because this number is measured in minutes and seconds, she'd have to deal with base 60 (time) and base 10 (percentages). Ugh.

What if she just noted the increased amount of time spent on her site? Simply subtracting these two values should do the trick and is a lot easier.

$$2{:}53 - 2{:}45 = 0{:}08$$

So on average, visitors are staying on the site for 8 seconds longer this month.

The bounce rate is already measured as a percentage. With a little bit of research, she finds that this number represents the percentage of visitors who "bounce" or leave the site quickly. It's pretty darned difficult to get a really low bounce rate, especially when you consider the number of people who visit a site just based on Google's recommendation. But getting this value as low as possible is a good idea.

If Emma finds the difference, she'll know by how many *percentage points* her bounce rate has dropped.

$$62.9\% - 59.4\% = 3.5$$

Emma's bounce rate has dropped. But it hasn't dropped by 3.5%. Nope, it dropped by 3.5 percentage points. There's a big difference, remember?

To calculate the percentage change of Emma's bounce rate, she has to subtract the old bounce rate from the new bounce rate and divide by the old bounce rate.

$$\left(59.4 - 62.9\right) \div 62.9$$

$$-3.5 \div 62.9$$

$$-0.0556... \approx -6\%$$

So her bounce rate dropped by 6% in one month. Not bad!

Sharing these numbers with a prospective editor or agent might not be the best idea at this point. But with this data, Emma can strategically plan how she'll broaden her platform. Within a few months, she should have some great numbers to share with the pros. And that just might land her a lucrative book contract.

Other numbers that publishers love

What if you don't have an impressive social media platform? Well, we've all got to start somewhere! If you're new to this whole process, you'll need to get creative. Just remember that your platform should be easily measured using numbers. Then start brainstorming. Here are some ideas to get you started.

Speaking engagements

Do you find yourself in front of a group talking about your book topic? Whether you're sharing your expertise with Girl Scout troops or a professional organization, figure out how to quantify those events. If an impressive number of people have heard you pontificate from the podium, use numbers. And if you haven't had that groundswell of followers yet, list your speaking engagements, including the number of people in the audience.

Media interviews

Have local or national media outlets asked for your expert opinion? List these. Separate them out into categories: television, radio, print, and online. Then estimate the number of people who might have seen or heard the interview. Television and radio stations can tell you the number of people who watch or listen to particular programs. For print publications, ask for circulation numbers (both subscription and newsstand sales). And for online media outlets—from podcasts to blog interviews— ask for the number of people who visited or downloaded.

Average number of comments per post

If you do have a blog—or blog for a third party, like a special-interest website or professional organization—

figure out the average number of comments per post. Some commenting systems will keep track of this for you, but others will not, so you may need to do some digging, and a little math, to get this number. Here's how: find the total comments for the last month and then divide by the number of posts. That will give you the average comments per post.

Newsletter subscribers

This is a particularly good idea for business owners or those working in the nonprofit sector. If your company or organization sends out a newsletter, include those totals in your proposal.

Membership in professional organizations

Only include this number if you can actually reach these members through a website, blog, newsletter, magazine, or conference. If so, make sure to mention the number of people in the membership rolls.

Math for Grownups

Chapter 7 — Spending the Advance

ONE REASON TO GO the traditional publishing route is the advance. When a publisher signs an author for a book deal, it often offers a lump sum of money before the book is even published.

And although these amounts have dwindled over the years, an advance can be incredibly useful for a variety of reasons. Some writers need that money for living expenses, especially if the research, reporting, and writing is particularly arduous. Other writers use the advance for costs incurred in the research and reporting itself—like traveling to interview sources or purchasing ingredients and equipment to test recipes.

How you spend that advance is completely up to you. But like any income, budgeting can help you make the most of your advance.

Andrea is a budding cookbook author. Her experience as a chef has provided the background she needs to develop and test recipes. And boy does she have a cool idea.

These days, families' eating habits are completely fractured. One kid has celiac disease and can't eat gluten. Another watched the latest documentary on factory farming and refuses to eat any animal products (yes, including cheese). With a demanding job, one parent barely makes it home before 7:00 p.m., pushing the family dinnertime way, way back. And of course, keeping the grocery bill in check is critical for the all-important budget, which is stretched to its limits.

A cookbook that would help families address these barriers is a perfect idea. In it, she'll describe how to customize go-to recipes for each eater, as well as ways to design low-cost meals that are easy to heat up. Perfect.

A great publisher thought so too. Andrea has been offered a pretty good advance: $6,000. She'll get $3,000 when the contract is signed and another $3,000 when the final draft is submitted.

But should Andrea shop this around a bit longer? She might be able to get a bigger advance. She's excited to have this nailed down already, though. Could $6,000 be enough?

She's got a few things to think about, so she makes a list:

1. Cost of ingredients for testing recipes
2. Cost of new equipment
3. Hourly rate for recipe development and testing, writing, and editing.

She thinks a little longer—does she need to include costs like office supplies, books for research, etc? Office supplies are part of her overhead, so that would be covered in her hourly rate. (More on that in the next section of this book, "Make More Money.") Most of the books she would use for research are either on her shelves or at the library. Any additional books she purchases are probably good for her personal bookshelves, so that's another overhead cost.

Andrea decides that this list is good enough. But how can she make a really smart decision?

In terms of total costs, equipment is going to be the smallest. She's got the basics—pots and pans, stove, oven, mixer, etc.—already in her kitchen. Anything more will likely be relatively inexpensive. She decides that $500 is a good bet for equipment costs.

Next up is the cost of ingredients. Once the recipe is developed, she'll need to test it 3 times before she can be sure it's good enough for the book. So if the ingredients for a recipe cost $25, she can expect to put out $75. With 50 recipes in the book, how much should she budget?

This is when Andrea is grateful to have a solid book proposal. In it, she included a variety of recipes that she's ready to adapt. The publisher gave her feedback on them, so she knows which ones will likely be in the final edition. (As long as testing goes well.) These recipes are also supposed to be relatively inexpensive. She estimates the cost for ingredients for each, and then finds the average. This gives her a rough estimate for the cost of each recipe: $12.43.

But she's got to test these bad boys 3 times. So she multiplies that average by 3:

$$12.43 \bullet 3 = \$37.29$$

Next, she multiplies by the number of recipes she expects to include in the book: 50.

$$37.29 \bullet 50 = \$1,864.50$$

"You know," Andrea thinks. "I might as well round that up to $2,000." This gives her a little cushion in case she needs it. (Not much, but a little.)

So far, Andrea has "spent" $2,500 of her advance—$500 on equipment and $2,000 on ingredients for recipe development and testing. That leaves her how much to cover her hourly rate?

$$\$6,000 - \$2,500 = \$3,500$$

Not much! But she knew this would happen. Andrea can now consider this in one of two ways. If she divides by her regular hourly rate of $100, she will know how long she should devote to cooking and writing. Or she could estimate the number of hours it will take to actually get the book to her editor and see what her hourly rate is.

She decides to consider both and starts with finding the time it should take to write the book given her $100 per hour rate and $3,500 fee.

$$\$3,500 \div \$100 = 35 \text{ hours}$$

Okay, so that's not going to happen, is it? Probably not. It'll take at least 8 hours to develop and test each recipe and there are 50 of them in the book. So recipe development alone will take:

$$8 \bullet 50 = 400$$

At $100 per hour, that amount alone is way more than the entire advance. And it doesn't include the time spent shopping, researching, and writing, not to mention failed attempts that can't even be included.

Fact is, Andrea's advance is not going to cover her time at her preferred hourly rate. No way.

So she decides to work backwards, estimating the time it takes to complete the cookbook and then finding the hourly rate. (This isn't going to be pretty, y'all. Brace yourselves.)

Shopping: 30 minutes per recipe (converted to 0.5 hours)
Recipe development and testing: 8 hours per recipe
Writing: 2 hours per recipe, plus 10 hours for
        introductions, etc.
Editing: 1 hour per recipe, plus 5 hours for introductions,
        etc.
Proofreading: 10 hours

With 50 recipes in the book, here are her totals:

Shopping: 0.5 • 50 = 25 hours
Recipe development and testing: 8 • 50 = 400 hours
Writing: (2 • 50) + 10 = 100 + 10 = 110 hours
Editing: (1 • 50) + 5 = 50 + 5 = 55 hours
Proofreading: 10 hours
TOTAL: 25 + 400 + 110 + 55 + 10 = 600 hours

With \$3,500 of her advance left, what will her hourly rate be? Andrea divides.

$$\$3,500 \div 600 = \$5.83 \text{ per hour}$$

Whoa. That's a far cry from her typical hourly rate. Sure, Andrea can hope to earn royalties, after she earns out—or sells enough books to cover her advance. But the reality is most books don't earn out. (Book writing sure can be a labor of love!)

Andrea could certainly shop her book around a little more. But she feels like given the current book market, this is not a bad deal, so she decides to take it.

Chapter 8 — Going It Alone

SELF-PUBLISHING HAS A LOT of benefits. There's no wooing of the publisher or agent, so writers can get to what they actually want to do: the writing. Authors have control over the entire process; from the book cover design to the length to the content itself.

Of course it's not all roses and hearts in the self-publishing arena. As any professional writer knows, just putting a book out there doesn't mean it will sell. Turns out that the writing might actually be the easiest part. And anyone who is considering independent publishing should look at the numbers before diving in. In particular, you want to know if the investment you make in self-publishing is going to pay off.

Michelle is thinking about self-publishing her next book. Her first baby—*Stand Up! My Comedic Journey*—was well received, but working with a traditional publisher was a bit disappointing. Besides, her current publisher didn't go for her next idea, *Writing Jokes that Sell*. It's a departure from the memoir she wrote last time, and the market is pretty targeted. A well-placed self-published book might be just the right thing.

Going through the process with a publisher the first time around, Michelle picked up a few things. First off, she needs to hire a good editor and copy editor. She's seen too many poorly edited self-published books to go that route. Second, she needs two designers: someone who can design the cover and someone who will create a design for the inside pages. (It's possible she can use a template for the book design, but she's not going on that assumption alone.) She'll also need to spend some money on promotion and marketing—like web design, printed promotional materials for events, and perhaps even advertising.

Next, she's got to make a big decision: print or e-book or both? Michelle knows that while an e-book is less expensive to produce, its cover price is much less. And while a print book costs more for the consumer, it also costs more to publish. She decides to consider both options and see what happens.

So how much is all of that going to cost? And how much can Michelle expect to earn from her book? She gets to work on some research.

For this bit of analysis, a spreadsheet works great. (Want to learn more about spreadsheets? Visit www.mathforgrownups.com and click "Writers" for special content on how to build spreadsheets with formulas.)

In a spreadsheet, Michelle can plug in the numbers she comes up with, along with a nifty formula or two, and *boom* she's got her answer. If she needs to adjust some of the costs, the final will adjust automatically. So she powers up her computer, opens up a spreadsheet, and starts creating her rows and columns:

	A	B
1	*Writing Jokes that Sell*	
2	publishing costs	
3		estimated cost
4	development editing	
5	copy editing	
6	cover design	
7	book design	
8	promotion	
9		
10	TOTAL	
11		

First, she considers her upfront—or development—costs. Going through the list, she considers each item one by one, doing research until she's got most everything completed. It's tough to figure out how much she'll spend on promotion, so she guesses at that amount. (She'll continue to spend on promotion after her book is published, using the profits from her book sales.) But the rest comes from really solid information gathered from smart people in the industry.

	A	B
1	*Writing Jokes that Sell*	
2	publishing costs	
3		estimated cost
4	development editing	$1,000
5	copy editing	$750
6	cover design	$300
7	book design	$300
8	promotion	$500
9		
10	TOTAL	$2,850
11		

Whether she chooses print or e-book or both, she's already in for nearly $3,000 on this venture. Now to find out the costs of publication itself.

Michelle has done a lot of research on this. She's decided to use an aggregator—a company that will format her text for a variety of different e-readers, as well as print on demand. The aggregator will also distribute her book to online stores and make it available to brick-and-mortar shops.

The cool thing about publishing a book this way is that she won't pay a dime upfront. Instead, the aggregator and the online bookseller will take a cut of each book sold. In other words, it works on commission.

The price of the book minus the commission is what is known as the royalty—and that's exactly what Michelle will earn. Royalties are different from place to place. Amazon.com offers a certain percentage, which differs from percentages offered by aggregators. And the royalty on a print on demand (POD) book is something completely different. That's because there are costs involved in printing a book.

Of course, royalties depend on something else that's pretty important—the price of the book. The higher the price of the book, the more Michelle will earn each time someone purchases the book. But pricing is always a balancing game. If she puts too high a price tag on her book, people might decide not to purchase it.

(There's a real science to pricing books, and we can't even scratch the surface here. It's a good idea for all independent writers to bone up on the pros and cons of different book-pricing strategies. Trusting your gut may not be the best option in this important decision.)

At this point Michelle needs to project her revenue, based on one aggregator's royalties. That information will help her finalize the prices for her book—both e-book and print options. To start with, she needs to choose a price for each and see what happens. Michelle does some research and decides on $12 for the paperback version and $4.99 for the e-book.

She decides to look at print options first. If she decides to print on demand (POD), online purchasers can buy her book one at a time. Plus, she has the option of ordering dozens at a time—a great option because she'd like to sell her books during her stand-up nights.

One aggregator she's considering has a super-handy calculator in its website. All Michelle needs to do is plug in some specs and out pop her royalty values. She makes the following selections:

> Black-and-white interior
> 5" x 8" trim size
> 150 pages
> $12 list price

Within seconds, she gets the following information:

> $4.55 royalty for each book sold on Amazon.com
> $6.95 royalty for each book sold on the aggregator's website
> $2.15 royalty on sales from other outlets on the aggregator's distribution list

Just like a traditional publisher, the aggregator that Michelle is considering offers author copies. These are copies that the author can purchase at a lower price than the cover price. Michelle can order as many of these as she'd like to sell at stand-up comedy shows and other events. Again, she makes her spec selections:

> Black-and-white interior
> 5" x 8" trim size
> 150 pages
> 50 copies

The calculator spits out the following information, which represents cost per book, not royalty:

$2.65 per book
$132.50 for 50 books
$23.00 for standard shipping

Michelle could certainly purchase 500 books to have on hand, but it might be less overwhelming to get them a few at a time. Just to see if she gets a break on the cost, she plugs in 500 books. Turns out she'll pay the same per book whether she buys 50 or 500 (aside from shipping costs).

In terms of print copies, Michelle is really pulling together a great plan. She can purchase multiple copies to sell in person, and she can have print copies available to buyers who would rather purchase books online—or for booksellers who want to have her book in their brick-and-mortar stores. (This isn't likely without a lot of marketing on her part, but it is available.)

Now for the e-book version. If Michelle prices her e-book at $4.99, she'll earn $3.24 per book sold on Amazon.com. The aggregator offers an 83% royalty, which nets Michelle $4.14 per book. Michelle knows that Amazon has about 50% of the e-book market, which leaves 50% for the aggregator's and other websites. So it makes sense to average these 2 values to get the net revenue per e-book.

$$(3.24 + 4.14) \div 2$$

$$7.38 \div 2$$

$$\$3.69$$

In other words, Michelle can expect to earn an average $3.69 per e-book sold. This isn't exact, but it's good enough for estimating how much revenue Michelle can expect.

Going back to her trusty spreadsheet, Michelle plugs in these figures and then does some additional calculations.

She wants to know the actual cost of publishing her book, if she sells 1,000 copies.

	A	B	C	D	E	F
1	Writing Jokes that Sell					
2	publishing costs					
3		estimated cost	price per book	NET per book	GROSS (1,000)	NET (1,000)
4	development editing	$1,000				
5	copy editing	$750				
6	cover design	$300				
7	book design	$300				
8	promotion	$500				
9						
10	TOTAL	$2,850				
11						
12	ebook	$0.00	$4.99	$3.69	$3,693	$843
13	print, Amazon	$0.00	$12.00	$4.55	$4,550	$1,700
14	print, aggregator's site	$0.00	$12.00	$6.95	$6,960	$4,110
15	print, other outlets	$0.00	$12.00	$2.15	$2,150	-$700
16	print, author's copies	$2.65	$12.00	$9.35	$9,350	$6,500

"NET per book" comes from the calculations she made above—the royalty she earns on each book sold, according to where and how it was sold. To find "GROSS (1,000)" or the amount she'd gross if she sells 1,000 books, she multiplies the NET per book by 1,000. Finally, the "NET (1,000)" is the GROSS (1,000) minus Michelle's development costs.

In the case of the e-book, print sales on Amazon, print sales on the aggregator's site, and print at other sites, this is pretty darned simple. She expects to spend $2,850 to produce the book. So she subtracts this amount from what she earns in selling 1,000 books.

$3,690 – $2,850 = $840

$4,550 – $2,850 = $1,700

$6,960 – $2,850 = $4,110

$2,150 – $2,850 = -$700

$9,350 – $2,850 = $6,500

To find her net for selling her books at events, she needs to subtract the cost to produce each book, as well as the total costs of book development. In the spreadsheet, that works out like this: she subtracts her cost for one book from the retail price of the book and then multiplies by 1,000. Finally, she subtracts the development costs from her net.

So, Michelle's revenue projections are kind of all over the place. Her research shows that she while she'll likely sell the most books on Amazon, she'll earn the least on those sales, both e-book and print. She can send folks to the aggregator's website to purchase a print copy—where she can earn the most money—but Amazon has at least 50% of the market, making it the most likely place people would find her book for sale. Print sales through other sellers are much lower, and it's most profitable for Michelle to sell her books herself.

Michelle can decide to sell only e-books or only print books. Or she can choose all of the options in her spreadsheet. If she does choose all of these, however, her development costs are only going to apply a single time. In other words, once she covers her development costs—by selling $2,850 worth of any kinds of books—all of her royalties are profit.

The one exception is this: if Michelle decides to spend additional funds on promotion, she's wise to consider those costs as coming from her profits. When she keeps a close eye on what she's spending to promote her book, versus what she's earning in sales, she'll likely earn more money in the long run.

At the very least, Michelle now has a great starting point with her spreadsheet. She can fiddle with the numbers a bit and see what the formulas spit out. She's on her way to a great plan for getting her book out to readers.

Part 3 — Make More Money

SURE, YOU LOVE TO write and report and research. But if you can't pay the bills, you might find yourself in a green apron, asking customers if they want skim or whole milk in their lattes. The magic formula is simple, but difficult to achieve. All of us freelancers are looking for ways to maximize the fun of writing *and* income.

But as any freelancer knows, there are a lot of hidden costs in this business. For example, magazine writers must write a detailed query to send to prospective publications before they even start writing the article. And some of these ideas never sell. Writers who work with nonprofits or for-profit companies must market themselves through a letter of introduction (LOI) or at special networking events. And authors often need to spend a great deal of time drafting a sharp proposal that will catch the eye of an agent and publishing house.

This upfront work takes time. But without a salary, freelancers are fronting this cost. In the business world, that's called overhead. It includes not only marketing time, but also materials, office space costs, and even insurance.

And we haven't even gotten to setting rates, determining whether a project is profitable, and dealing with issues like cash flow.

Numbers, numbers, numbers, right? The math involved in this is really basic. In fact, if you read through the first section of this book, you got a great review. Most of it requires a good understanding of percentages, arithmetic, and logic. If you can find the tip on a restaurant bill—in your head or with a calculator—you've got this.

Unfortunately, some freelance writers ignore this math. Maybe it's because they're scared or feeling not so confident. Maybe it's because they'd rather fly by the seat of their pants or trust their guts.

Whatever the reason, not doing the math is a dangerous proposition. First off, you can end up on the losing end of a project. Secondly, you can miss out on opportunities for growth.

Now, I'm the first to admit that not everyone wants to earn a six- or seven-figure salary (or more). But one thing is for sure: if we don't have goals, it's less likely we'll reach higher heights. Whatever you want to do—make more money, work fewer hours, get published in your dream publication, write a *New York Times* Best Seller—math can help. Promise.

But if you need a little more convincing or a review of how this math is applied, this section is for you. In it, you'll learn how to calculate your hourly rate and turn it into a project fee. You'll also get the scoop on diversifying your client base, so that you're not vulnerable to big changes in the industry. And you'll learn how to set goals so that you can keep an eye on the books—while having much more fun writing.

Now, you might say, "I have an accountant for all of this!" Terrific! But you're selling you and your business short if you have to run all of these things past someone else. I'll show you how to do these quick calculations so you can make quick decisions.

Ready? Let's do some make-more-money math.

Chapter 9 — What's Your Hourly?

FREELANCE WRITERS GET PAID in a variety of different ways—by the project, by the word, by the hour, by the number of pageviews, and by a percentage of sales. And let me tell you, those figures can run the gamut, from $3 per word down to $0.25 per word or from $800 per blog post to $5.

How can you manage those crazy inconsistencies?

For many longtime freelancers, an hourly rate can help. If you know how much you need to earn per hour to cover your overhead and expenses and allow for the occasional splurge on good chocolate, and you know (or can make a good guess) how long a given project will take, you can calculate whether what you will earn for said project (however it's paid) will be worth it.

But how do you set your hourly rate?

If you set it too high, you could be turning down a lot of work and end up not making enough to live on. And if you set it too low, you could wind up feeling resentful and abused.

The hourly rate is determined by two parts: your annual salary and the number of hours you work in a year. In a 9-to-5 job, your employer sets these parameters. As an independent writer, you make those decisions—either consciously or unconsciously.

(Seriously, you *do* have an hourly rate, even if you don't know it. Wouldn't it make more sense to be proactive about it, instead of letting the chips fall where they may?)

In the end, setting an hourly rate has more to do with *your* financial needs and goals than clients' ability to pay. As an initial exercise, why not figure out what you would *like* to earn and see if it makes sense?

Kelly is about to do just that. She's new to freelancing but an old friend of journalism. She recently left her job as a newspaper business reporter, and she's got a few national clips under her belt. She knows what market she wants to target and has the contacts. But first she wants to spend some time thinking about how she's going to earn a living as a freelancer.

As a journalist, she earned $52,250 per year. She'd like to match that this first year as a freelancer. Kelly knows she'll take on some expenses that her employer used to pay—like health insurance and the employer's share of taxes. But she has been living below her means for quite some time. Looking at her budget, she figures she can manage those on her newspaper salary—as long as she watches her incidental spending.

So Kelly's got the first part of the hourly-rate equation: her annual salary. Now for the hours she wants to work. One reason she decided to become a freelancer is to work fewer hours. But again, she's just starting out, so she decides to go for a regular working schedule: 40 hours a week, 48 weeks a year. She figures a total of four weeks off—for vacations, holidays, and sick days—makes sense.

Wait a minute, Kelly thinks. When she was a reporter, she was paid a salary that covered her vacation time, holidays, and sick leave. Shouldn't she do the same? Of course. So she makes sure to include 52 weeks, when she finds the total hours worked in a year:

40 hours • 52 weeks = 2,080 hours

There she has it—the two parts needed to find her hourly rate.

Annual salary = $52,250

Hours worked in a year = 2,080

To find the hourly rate, all she needs to do is divide. She remembers that she's finding the dollars per hour, so she divides the salary by the number of hours.

52,250 ÷ 2,080 = $25.12

Really? That's a pretty low hourly rate. She knows other freelancers who are charging $100 to $150 per hour. Why the huge difference?

It's simple, really. At her newspaper job, Kelly had guaranteed work. She was paid for writing, not for tracking down assignments, bookkeeping, messing around with computer glitches or managing a website. These jobs were outsourced to other people in the company. But even if Kelly outsources them as an independent writer, she's got to foot the bill.

Sigh. This is more complicated than she thought.

What Kelly didn't consider is overhead. She's no longer just a writer but also a small business owner. And those costs have to be paid somehow.

But how can she get those into her hourly rate? She could just shoot for $100 per hour, but Kelly wants to dig a little deeper.

Instead, Kelly decides to look at how she expects to spend her time. An extremely detailed and organized person, Kelly has given this a lot of thought. In this first year, she figures she'll spend about one-half of her time (50%) on marketing, which includes managing her website and social media, researching new media outlets, coming up with story ideas, and pitching.

After that, she allows herself a quarter of her time (25%) for administrative stuff, like bookkeeping, organizing, and dealing with computer issues. That leaves 25% of her time for actual reporting, writing, and editing.

This is a really rough estimate, Kelly knows. But it's a good-enough starting point. So she does the math. Her goal is to find out how many actual writing hours she has in a week. Then she can determine her hourly rate based on those. That rate will cover her overhead.

Marketing: $0.5 \cdot 2{,}080 = 1{,}040$

Administrative: $0.25 \cdot 2{,}080 = 520$

Writing: $0.25 \cdot 2{,}080 = 520$

So she's estimated that she'll spend 520 hours a year on actual reporting and writing—that's 10 hours a week, if she's working a regular 40-hour work week. It seems pretty low, but Kelly's just getting started. In order to get the gigs she wants, she'll spend far more time landing assignments than actually doing them.

To find her hourly rate, she divides what she wants to earn by the total writing hours:

$\$52{,}250 \cdot 520 = \$100.48$

Well, how about that! It's darned close to the rate she was told to charge. And it sounds about right for a new freelancer. Kelly settles on $100 per hour.

A different path

You can find your hourly rate in different ways. For example, you can add up your total costs—from printer ink to website development to taxes to insurance—and add those to an estimated salary for a writer in your field. Then you can divide by 2,080—or the number of work hours in a year (assuming a 40-hour workweek).

(business overhead + typical writer salary) $\div$ 2,080

Another option is to add all of your business overhead (including taxes) to personal overhead (like housing, food, entertainment, and student loans) and then divide by 2,080.

(business overhead + personal overhead) ÷ 2,080

Any calculation of an hourly rate needs to account for the time you spend on work that you can't bill, such as working on your website, researching publications, pitching new markets, and billing clients. That's why Kelly made the calculation the way she did. She assumed that she would only be able to bill 520 hours of work this year.

Turns out, her instinct was pretty good. Many new freelancers underestimate their hourly rates. Somewhere between $75 and $150 is a pretty good range—depending on where you live.

But don't just pick a number out of the air. Do a little bit of figuring. When you have a realistic hourly rate in mind for each project, you'll find it much, much easier to decide whether to take a project and set financial goals that you can actually reach.

Chapter 10 — Putting on a Price Tag

SO YOU'VE GOT YOUR hourly rate. Great! Now you've got to figure out how to use it. As I mentioned earlier, there are a variety of ways freelancers are paid for their writing:

1. By the word
2. By the hour
3. By the project
4. By the pageview
5. By a percentage of sales
6. A combination of these

(Before getting started, here's a quick note about number 4. When writers are paid by the pageview—sometimes called "per click"—they're paid a certain amount for every person who clicks on an online story the writer has written. This payment is usually parsed out by a certain number of clicks—like $1 per 1,000 clicks. While this may be a fair way to be paid, it can also be a real red flag. When writers are asked to publicize pieces that they have written, they're shifting more of their time into marketing, rather than writing. Marketing is good, but do you really want to spend your time promoting someone else's website? As always, there are reasons to go this route. Just make sure you're making them consciously, rather than simply signing up out of excitement or desperation.)

Magazine writers totally get the pay-per-word option. This is standard for most consumer pubs. This rate has also been flat for more than 30 years. Yes, more than 30 years. Many magazines are paying the same per-word rate now that they were offering in the 1970s. Crazy, huh?

Companies and nonprofits are more likely to offer a per-hour rate. In this case, you might be considered a subcontractor. The client may want you to suggest the hourly rate, or it may have a rate in mind.

In my opinion, project fees are the best options for freelancers. Think about it. How do you feel about telling your friends that you charge $150 per hour for your writing? If they're typical 9-to-5ers, they may have no concept of overhead or have no clue what it means to pay your own taxes. In addition, they may not have a clue how much their own hourly rate is. Clients feel the same way.

And there are other reasons to think of your work on a project basis. If you know how much time it takes you to complete a project, you can find out if a per-word rate will get you to your desired hourly rate. See? The flexibility of math! Let's take a look.

Sandy has been writing for national magazines for years. She is tickled pink to have finally broken in to her dream publication. They pay a whopping $2.50 per word, but she's heard the editing process is tough. "Tough" translates to "a lot more time than Sandy's other assignments."

Even though it's her dream pub, Sandy knows she needs to do the math before saying yes.

The assignment is an 800-word story that will require three interviews. Right away, she can calculate the amount she would be paid.

$$\$2.50 \bullet 800 = \$2,000$$

(If you're new to the biz, you may not be aware of this. It's pretty rare that a publication will pay for exactly the number of words you turn in. Instead the rate is based on the assigned words. Sometimes this works in the writer's favor. If Sandy turns in 789 words, she still earns $800. And sometimes it doesn't. If she turns in 836 words, she still earns $800.)

As far as magazine writing goes, $2,000 is not bad. But how does it work out in terms of her hourly rate?

Here, Sandy has a couple of choices. She can estimate the number of hours it will take her to write the story and then figure out the hourly rate the story offers. Or she can divide $2,000 by her hourly rate of $125 to find the number of hours she can spend on the story.

The first option makes more sense to Sandy, so she tries it out. In her mind, there are four parts to reporting and writing a story: interviews, research, writing, and editing. Given her topic, the length of the story, and the reputation of the publication, she comes up with these estimates:

Interviews: 3 hours
Research: 2 hours
Writing: 6 hours
Editing: 4 hours

That's a total of 15 hours. Now to divide the total story fee by 15 hours.

$$\$2,000 \div 15 = \$133.33$$

Not bad at all! In fact, Sandy's set to make a little more than her hourly rate on this story.

One great benefit of this process is that Sandy now has goals for managing her time. If she's able to cut her interview time down to 2 hours, she can spend more time writing. But if her research time goes way over 2 hours, she should try to make up for it in writing.

In fact, Sandy overestimated the time spent on interviews and writing. That gives her a little cushion, if the words just won't get on the page or if editing gets out of hand. At the same time, that cushion allows for thinking time—when she's washing dishes, exercising, or in the shower (where the best ideas appear). She can count that time if she wants or she can just consider it part of her writing process and roll it into one of the other categories.

Developing project fees for each assignment is an extremely powerful tool that can keep you on track with your hourly rate and goals. But it can take years to really hone in on your timing. Everyone—even the most experienced writers—underestimates the time it takes to conduct interviews, write stories, and edit. You do yourself a big favor by tracking your time carefully.

The key is to stay on top of these numbers, if you want to actually make a good living. Accounting programs and online calculators can help, but having a good sense of your rates and the time it takes for you to complete projects is paramount.

Getcher "ohm" on

When it comes to your hourly rate, you really need to take a page from the great yogis out there—be flexible. Knowing your hourly rate doesn't mean you're *always* going to get it. And here's another little tidbit: it doesn't mean you have to turn away work that doesn't meet your hourly rate.

Like a budget, your hourly rate is merely a guideline for you follow. There are times when you'll decide to work for free. There are times when promotion is going to be a great benefit of a low-paying gig. And there are times when you decide that you should gain some experience with lower-paying gigs before stepping up to the plate with the big guys.

But when you make these compromises, you absolutely have to balance them out somehow. Your work should be like one of those mobiles that hang over baby's cribs or in fancy museums. In order for the mobile to hang properly, each piece must be balanced. When one is too light, the others will be too heavy and vice versa.

So if you decide to blog for Huffington Post, which doesn't pay bloggers at all, you need to figure out how you'll balance that time (and lost revenue) somewhere else. If you don't, you'll miss your income goals and could run into some big cash-flow problems.

Let's say it takes about three hours for you to write and edit a HuffPo blog post. If your hourly rate is $150, you need to make up $450 somewhere else. That dough can come from a variety of places. If you have a monetized blog, you could generate more income from the traffic that visits your site from HuffPo. Or you could generate more book sales with the exposure.

But in all likelihood those things ain't gonna happen. First, you can't guarantee that your post, much less your blurb with your book title and link at the bottom, will be seen by many people. Secondly, it's awfully difficult determine whether or not a spike in your site traffic or book sales is due to your writing on Arianna Huffington's site.

And think about the long-term effects of taking on unpaid work. To build an audience—even at HuffPo—takes a lot of work. You might need to write four to six posts per month. That's $1,800 to $2,700 per month you're losing and have to make up elsewhere!

In these situations, it can be a good idea to plop that time into your marketing hours and not count it as billable time. Then track what happens. If you get consistent increases in site traffic, subscriptions, or book sales, that free gig could be worth it. But if you don't, well, you might be better off spending that $450 per post somewhere else or finding a paying client.

The point is this: just because you've done the math doesn't mean you must live and die by the numbers. Math allows you to look at the bigger picture, to quantify your career, and to measure your successes and failures. What you're looking for is a balance, an average hourly rate that, by and large, meets your needs.

Namaste.

Chapter 11 — Go-o-o-o-a-l!

IF THERE'S ONE THING that separates the amazing business owners from the not-so-successful, it's this: folks on the top set measurable goals and track their progress.

No kidding. When you know what you're shooting for—and make it easy to determine when you've made it—you're more likely to get where you want to be.

It's like taking a trip across the country. If you want to arrive to the opposite coast by a certain date, you'd best make some plans for how long you're going to stay at destinations along the way. Otherwise, you'll end up with 7,000 miles to travel in the last day and only a Dodge Dart to get you there.

So say you'd like to make a six-figure income. It's not likely that you're going to achieve that on day one of your freelance writing career. But if you set some smaller goals, you could get there within a few years—or sooner.

Some of your goals are going to be things that are harder to track—like breaking into a dream publication or interviewing your hero for a story or landing a book deal. But when it comes to income, you can—and should—track how things are going.

Of course math can help with this.

Marijke has been freelancing for a while, but lately, she's noticed that her income is flat. While she doesn't need to see her income rise steadily each and every year, she does need a little extra cash these days. With three kids in college and retirement just around the corner (hopefully), she needs an influx of cash to make sure the second half of her life is as amazing as the first.

Smart girl, that Marijke. She's always had goals and followed them carefully. But she decides to put a little more oomph into that part of her business planning this year.

First, Marijke decides to set a larger annual income goal. For about three years, she's been hovering around $125,000 per year. Not bad. But what would happen if she stretched a little?

Of course, the amount Marijke should stretch is tricky. She decides to break things down, analyzing her current income as she goes. Then she can set two goals—one that she imagines will be pretty easy to reach and a "bonus goal," which is more of a stretch. She'll make financial plans using the lower goal and hopes that the higher one will encourage her to work harder and smarter.

Based on a $125,000-per-year income, Marijke first finds her monthly gross income. This math is super-simple. She just divides by 12.

$$\$125,000 \div 12 = \$10,416.68$$

She's been tracking her monthly income, so this is no surprise. To make the math easier, she rounds to the nearest 500 to get $10,500.

So right now, she's averaging $10,500 per month. By tightening up her negotiating process, shooting for higher-paying markets, and landing a few, simple copywriting gigs, Marijke believes she can bring in an extra $500 per month. That brings her new goal to $11,000.

So what is that annually? Easy calculation:

$$\$11,000 \bullet 12 = \$132,000$$

Just for the heck of it, Marijke decides to calculate the percentage change from her current income to her goal income. She wants to know, is it within the typical range of raises offered to non-freelancers—or between 1% and 4%? With this information, she'll have a better sense of whether her goal is reasonable.

Having just relearned how to find percentage change, Marijke goes slowly. The income went up, so she should see a positive change. She thinks carefully about the formula and jots it down:

(new value – old value) ÷ old value

Next, she substitutes and solves.

$$\left(132,000-125,000\right) \div 125,000$$

$$7,000 \div 125,000$$

$$0.056$$

So if she rounds her answer and changes it to a percentage, she gets 6%. The typical raise for traditional workers is 1% to 4%, so she's a little bit high. It's worth it to try once more.

Since she felt pretty good about a $500 monthly increase in income, which translates to $132,000 per year, she decides to shoot for $130,000 per year. If that gets her between 3% and 5% increase, she'll figure out the monthly difference.

$$\left(130,000-125,000\right) \div 125,000$$

$$5,000 \div 125,000$$

$$0.04 = 4\%$$

Well looky there! In less than 15 minutes, Marijke has hit upon a great number. Her safer annual goal is $130,000. That's only an extra $5,000 per year, which seems like not much. But it is very realistic, since it falls within typical income increase standards.

Very quickly, Marijke translates this new annual goal to a monthly one. (This is important. When she books assignments each month, she needs to know if she's on track for meeting her annual goal or not.)

$$\$130,000 \div 12 = \$10,833.33$$

Remember, her original monthly gross income was $10,416.68. Subtracting her goal from her actual income, she finds that she needs to bring in an extra $416.65.

Now, because these are goals, Marijke doesn't need to follow these numbers to a T. For example, she can round her new monthly goal to $415 or $420 or $425.

Marijke is amazed at how a little jump in her monthly goal can boost her income *and* make her feel like she's really growing. But what about that bonus goal?

The bonus goal is important to Marijke because it really helps her stretch. Even if she reaches it only a few months out of the year, she's already made more money than she expected. It's neat little psychological trick that keeps Marijke on her marketing and negotiating toes.

Marijke decides that if she can earn an extra $5,000 per year, she could shoot for another $10,000 per year as her bonus goal. That means she'd gross $135,000 annually. What does that shake out to be monthly? And how does that compare to her current gross monthly income?

$$\$135,000 \div 12 = \$11,250$$

$$\$11,250 - \$10,416.68 = \$833.32$$

So to meet this bonus goal, Marijke would need to bring in an additional $833.32, rather than $416.65—twice as much. It's definitely a stretch, but that's what makes it a good bonus goal.

With all of this figuring behind her, Marijke feels pretty darned good about what she can accomplish next year. The extra monthly income could come from raising her rates slightly or being a little tougher negotiating payments or even monetizing her blog. Those possibilities are endless, and that stuff is far more interesting to think about than the actual math.

Chapter 12 — Laying a Sound Foundation

THE ECONOMIC DOWNTURN HAS taught independent workers, a really important lesson: diversify, diversify, diversify. It's been a long, long time since the publishing industry has seen such a shift in the way it does business. The Internet has certainly played a role. At the same time, a poor economy forced consumer, custom, and trade publications to slim down or close up shop.

That left a lot of freelance writers wringing their hands.

It also forced lots of traditionally employed writers and editors into the freelance business. Suddenly there was even more competition for slim pickins on the glossy pages of famous magazines. It was tough.

But there was a group of writers who not only survived but thrived. These smart folks knew it was critical to diversify their client base. That way, if the bottom fell out of the custom publishing market, they could be buoyed by their contacts in, say, the nonprofit sector.

There are so many different ways to diversify your income. The goal is to make sure that your income comes from enough different kinds of sources that if one type falls apart—or you become disinterested in doing it—you have something else to build on, at least temporarily.

But it's not enough to have a gut feeling about this. Your mind can completely fool you. Instead, it makes sense to set goals for diversification and track them.

One way to do this is to assign a percentage to each revenue stream. But for these to really work, they should be realistic, not numbers plucked from the clear blue sky. In addition, if all of your revenue is active—that is, it requires actual work hours to get it accomplished—these income percentages should roughly mirror the percentage of time you spend on each revenue stream.

Got it? Well then, let's look at an example.

Since starting her freelance business, Linda has been a rock star consumer magazine writer. Her byline has appeared in all of the biggies, but when the bottom fell out of the magazine industry, she saw the writing on the wall. She can still land a few of these assignments a year, but she needs to focus on other ways to make a living as a writer.

Her options are endless. She can parlay her magazine-writing experience into some gigs with trade or custom publications. She can tap the corporate market, pitching herself as a writer who can make white papers and annual reports come alive. Or she can get cozy with some nonprofits and universities, offering her services as a copywriter for educational and donation materials. The list goes on and on.

Linda already has contacts with a few businesses, so she puts that on the list as a definite. Then she spends a few hours researching the trade and custom publishing industry. These assignments are much more like the magazine writing she's been doing and the industry wasn't hit as hard as consumer magazines were, so that's a very viable market for her.

Linda ends up with a list of markets to weave into a set of diversified revenue streams:

Consumer publications
Custom publications
Trade publications
Businesses and corporations

Now to set percentages for each of these. Because she's starting from scratch, this is going to be a bit of a crapshoot. But she considers her current contacts and the promise of each industry in her decision.

She'd like to really scale down her dependence on assignments from consumer pubs.

Consumer publications   10%
Custom publications
Trade publications
Businesses

But what about the others? The math is really simple here, really. The total must be 100%, and with 10% already taken, she's left with 90% to divide between the remaining three categories. (Told you it was simple.) She could divvy it up into equal percentages, by dividing 90% by 3.

Consumer publications   10%
Custom publications   30%
Trade publications   30%
Businesses   30%

Over time, Linda can fine-tune her diversification percentages, but based on what she knows about each of these markets, this is a fine place to start.

And she's off! Linda puts the pedal to the metal in her marketing efforts, sending out letters of introductions, researching new clients who fit her portfolio, and networking with other writers and anyone else who might have an interesting lead for her.

After a year, Linda has some great numbers to look at. Changing paths was tough, but with some extra effort, Linda made some great progress, earning $82,000—just about as much as she did the previous year. (Next year, she hopes to pull back a little bit and settle into an easier schedule.)

Her records show these revenues for the year:

Consumer publications	$7,380
Custom publications	$26,240
Trade publications	$31,447
Businesses	$16,933

But how do those mesh with her diversification goals? She notices that the last three are not equal, even though her goal for each was 30%. So there is definitely something off with these. To find out what, Linda needs to find the percentage of each revenue stream. And to do that, she merely needs to divide each by her total income, $82,000.

Consumer publications	$7,380	9%
Custom publications	$26,240	32%
Trade publications	$31,447	38%
Businesses	$16,933	21%

With the percentages, it's very easy to see how her actual revenue for each is not quite what she had expected. She's fine with that, though. None of the revenue streams are even close to 50%, so she's not putting too many eggs in one basket. This was the whole point of the change.

So Linda has two choices: she can adjust her goals to better match these percentages or she can keep her 10-30-30-30 goals and resolve to pull up her business writing revenue. Regardless of what she decides, the math helped her get to this spot.

Diversification 2.0

Another really important consideration in diversification is the clients themselves. (And by clients, I mean magazines, businesses, book buyers—anyone who purchases your services or your products.) It's a common experience. We find a terrific client or meet an amazing editor, and we're ready to get all exclusive.

This is no big deal if you want a regular staff job. But if you want to keep freelancing, this mistake could be costly.

I know, because it's happened to just about all of us. The editor we love is fired, laid off, decides to switch careers, or just plain tires of us. We're pushed aside for the next big thing, and there's a huge hole in our income.

So along with maintaining diverse revenue streams, it's important to keep diverse client bases. This means keeping track of the percentage of your income that comes from each client. These don't need to be teeny-tiny, but they should be easily replaceable, if you lose that client. Replacing 4% of your income is a heck of a lot easier than replacing 40%. Right? Right.

Epilogue

BEFORE SCRIBBLING *QED* ON this book about using math to become an accomplished, published, and wealthy writer, I leave you with this: while math may not be easy, it *is* worthwhile. And now that you have a book that can walk you through some of the tougher challenges, you can certainly achieve far more than you ever thought possible.

But this book could only scratch the surface. If you have further questions, please join me and other writers at www.mathforgrownups.com. Click on the writers section to find more in-depth descriptions of the math that can help you in your career, plus a few options for practicing. And if you don't see what you're looking for, just drop me at a line or jump into the comments section to ask a question.

(Ummm... what the heck is *QED*? Why, that's the acronym for the Latin phrase, *quod erat demonstrandum*. It means, "that which had to be demonstrated" and appears at the end of math proofs. See? I told you math was a whole lot like writing. Believe me now?)

About the Author

WHEN PEOPLE LEARN THAT Laura Laing has a degree in mathematics, it's always the same reaction: Widening eyes change to a puzzled look and then, "But aren't you a writer?" Laura contends that writing great non-fiction is not much different from proving $a(b+c) = ab + ac$, except she gets to use words that are a whole lot more fun.

Yes, she taught high school math, but after leaving the classroom, Laura began applying her verbal skills to her career. In her 12 years as a freelance writer, she's written for a variety of regional and national publications, including *Parade, Parents,* and Southwest Airline's *Spirit* magazine. Her cover story for *City Paper*, about the oldest gay bar in Charm City, earned her a 2008 A.D. Emmart Award honorable mention. Her corporate and non-profit clients include Kennedy Krieger Institute, University of Baltimore, the Association for University Centers on Disabilities, and American Pool. She's also written online and print content for middle and high school mathematics curriculum.

For Laura, writing is teaching. And in July 2011, she published her first book, *Math for Grownups.* A funny, accessible and practical book, it's designed for readers who are afraid of math or think there is something called the math gene — and they don't have it. Read an excerpt from *Math for Grownups* on the next page.

Excerpt from Math for Grownups

THE NUMBERS GAME PLAYS a starring role in almost every part of daily life, from making dinner to planning a weekend getaway. Heck, you need math to order a pizza.

Even people who don't sweat math problems—namely, mathematicians!—sometimes have trouble with the calculations we face in everyday life. So it's no wonder that those of us who have already forgotten what we learned in high school (or, worse, never liked math and didn't do well in it) can sometimes stare at a math problem and not have the faintest idea what to do about it.

But luckily you don't have to be Steven Hawking (or even be proficient with a scientific calculator) to use math in ordinary situations. Remember, it's only a tool. (And it's not even one that requires safety glasses or special training so you won't cut off the tip of your left index finger.) It's a language that describes how our world fits together. Math enables us to make predictions and quick decisions. Math helps us feel powerful and confident.

Here's the honest truth: adding fractions is no harder than signing up for the office football pool or buying airline tickets online or remembering how to create a folder on your computer desktop. It may just *seem* more challenging. The truth is that very few people in the world can't do math. You are not one of them. Here's the thing—most math doesn't require you to remember how to find the slope of a line (or even to remember what slope is). The everyday stuff is a combination of basic arithmetic and your innate understanding of how to speak the language of numbers, shapes, and measurements.

Yep, innate. You were born with curiosity about the world around you. Math is just one way to describe that world. And, like it or not, it's a pretty important way.

So unless you don't care what's in your bank account or whether your new elliptical machine will fit through the door of your exercise room, you're going to have to do some math.

And you might as well think you're good at it, right? (Because, guess what? You are.)

You don't have to know calculus to figure out how to lower your monthly mortgage payments. You don't have to remember the Pythagorean Theorem to lose a few extra pounds. And you don't have to do long division in your head to buy paint for your new house.

You do need to have an open mind and a sense of humor. After all, it's only math.

www.ingramcontent.com/pod-product-compliance
Lightning Source LLC
Chambersburg PA
CBHW060900280326
41934CB00007B/1130